Aromatic C-nitroso Compounds

Hrvoj Vančik

Aromatic C-nitroso Compounds

Springer

Hrvoj Vančik
Faculty of Science and Mathematics
Department of Chemistry
University of Zagreb
Zagreb, Croatia

ISBN 978-94-007-6336-4 ISBN 978-94-007-6337-1 (eBook)
DOI 10.1007/978-94-007-6337-1
Springer Dordrecht Heidelberg New York London

Library of Congress Control Number: 2013933879

Printed on acid-free paper

Springer is part of Springer Science+Business Media (www.springer.com)

Preface

It is almost impossible to explain how a scientist who is engaged in the basic science creates his field of research. This personal process cannot be comprised within any of the formal concepts of philosophy of science. It could just appear that, during the work, an extra observation discovers a new phenomenon so interesting and intriguing that it directs the researcher to investigate in quite a different course.

My previous scientific interest in the structure and vibrational spectroscopy of carbocations in solid state was developed in such a direction, that the next step must include also a study of other organic cations, which could appear as reactive intermediates in organic reactions. The first candidate was the nitrenium cation, an extremely elusive molecular species. The task was difficult, but it is known that doing the serious science means a selection of difficult problems to deal with. On the way to the isolation of nitrenium cations, the study of appropriate precursors was necessary. One of them is nitrosylchloride, which is a reactive molecule that easily adds to the carbon-carbon double bond forming the alpha-chloro nitroso compounds. Since our approach has been based on the reactions of intermediates in the solid state, one of the molecules interesting for study was 3-chloro-2-nitrosonorbornane. Following the standard procedure in the solid-state chemistry, it happened that we had irradiated the crystals of such a norbornane derivative by UV light under the cryogenic conditions. Surprisingly, the crystals changed color from yellowish to deep blue. By warming the sample the color disappeared. Successive change in color induced by irradiation and warming could be repeated several times. Discovery of this photochromic-thermochromic effect in the solid-state seemed so interesting that it had redirected our research program to the new field of nitroso compounds. The observed photodissociation of nitroso dimers to monomers and their redimarization in crystals or in polycrystalline environments seemed to us as an opportunity to use these molecules as models by which basic solid-state reaction mechanisms could be studied. Such a way our new research field was opened. Later, after the experience in nitroso chemistry, aromatic derivatives emerged as especially interesting because their chemistry can serve as modeling system not only for the solid-state reactions, but also for a series of the chemical phenomena and concepts.

After publication of our basic works, I had the honor to receive a letter from Prof. Brian G. Gowenlock, who is a nestor in the chemistry of nitroso compounds, and who has inspired me to continue in investigating these fascinating molecules. Our correspondence was for me a strong intellectual impulse for continuing our study in this direction, and I take here the opportunity to thank Prof. Gowenlock for his suggestions, support and inspiration.

This book is also a result of the idea to prepare a manuscript about my research, which I have received from the Editor, Springer-Verlag. After some period of consultations I accepted to write the book about the chemistry of C-nitroso compounds with the intention to comprise not only our recent work, but also the wider aspects, such as spectroscopy, typical reactions, or reactions of biological interest.

The work in this form is the result of all of my collaborators in the research group, as well as of my colleagues who, by their suggestions and comments, noticeably improved the manuscript. I thank Prof. Zlatko Mihalić, Prof. Srđanka Tomić-Pisarović, Dr. Ivana Biljan, Dr. Srđan Milovac, Ms. Katarina Varga, my students Ana Maganjić and Ivan Šolić, and especially librarians Branka Maravić and Zdenka Kuri. The research that was the base for a great part of this work has been financially supported by the Ministry of Science, Education and Sports of the Republic of Croatia (Grant 119-1191342-1334). The support is gratefully acknowledged.

Zagreb, Croatia
10 November 2012

Hrvoj Vančik

Contents

Introduction

Molecular species that appear in nature could be classified starting from many different aspects. One of them is on the basis of their stability and possible persistence under different conditions. Looking from the potential energy hypersurface aspect, the molecule-like structures appear either in the energy minimum (stable molecules), or close to the saddle points on potential energy surface (transition states). Among the stable substances, some of them can survive long time if kept closed in bottles, but others, like the reactive intermediates for instance, appear to be relatively short-lived structures. If we go further in the more sophisticated resolution of the stability of molecular species, we can find molecules that are more stable than reactive intermediates, but still enough elusive that their conservation requires a lot of care. Nitroso compounds with NO group bounded to the C-atom, and especially aromatic ones, just reveal such properties. As we will demonstrate later in this book, aromatic C-nitroso compounds seem occasionally chimeric because of their dynamic, which can be recognized as equilibrium between monomeric and dimeric structures. Even more, aromatic molecules with more than one nitroso group can form polymers by a polymerization process that does not require special conditions for initiation. Brian G. Gowenlock proposed the intriguing idea of "living polymers".

The ability of nitroso molecules to form dimers is of wider mechanistic interest because the study of such dimerizations opens also another philosophy, the use of nitroso molecules as models for studying other chemical phenomena of more general importance. Use of the behavior of aromatic nitroso compounds as such models is mostly based on their strong structure-dependent dimerizations to azodioxides. Because the azodioxide bond between two nitrogen atoms is much weaker than normal covalent bond but stronger than hydrogen bond, shifting the equilibrium either to the dimer or monomer side could be fine-tuned by varying structure or environmental conditions. For instance, while in solution most of substituted nitrosobenzenes exist as monomers, they readily dimerize in solid state. Or, while *p*-nitronitrosobenzene forms azodioxides with most of the nitroso functionalized partners, the *p*-aminonitrosobenzene remains monomer in all the known examples. Investigations in this line can be of general scientific interest because such substituted nitrosoaromatic molecules can serve as excellent models for studying intermolecular selectivity, the property that is crucial in studying supramolecular design.

Since some combinations of the nitrosoaromatic molecules form dimers only in crystals, this molecular model opens opportunities to study some of the basic principles of molecular self-organization, and today especially important crystal engineering. Aggregation of molecules on the organized surfaces (for instance self-assembled monolayers, SAM) can also cause formation of the azodioxide bond. The phenomenon opens possibilities of deeper and very fundamental insights into the problem of self-organization of molecules.

In our research in this field, we have focused on the phenomenon that a series of freshly sublimed nitrosobenzenes appear as monomers, but at a slightly elevated temperature dimerization takes place. Measuring the rate of this process enables the study of the phenomena related to solid-state reaction mechanisms. Aromatic C-nitroso compounds seem to be also the molecular models, which could help in developing rational concepts of solid-state chemical processes. Even more, the discovery that nitroso dimers in crystals undergo photodissociation has at least two perspectives, the study of topochemical effects in the solid-state chemical reactivity, and the possibility to apply this effect in the molecular OFF-ON switches in molecular electronics. Additionally, since monomers and dimers differ in color, monomers are blue or green and dimers yellowish or colorless, such a photochemically induced solid-state reaction can be visualized, and the effect represents a new example of photochromism.

Chemical behavior of nitroso compounds is extraordinary. First of all, in dimerization, nitroso molecules show their ambivalent nature. During dimerization of a pair of molecules, one of nitroso partners behaves as the nucleophile, but the other plays the role of the target for the nucleophile attachment. Which one will take which role probably depends on the spatial intermolecular orientation.

Reactivity of the nitroso function could be compared with the behavior of the analogous carbonyl group. Nucleophilic addition to nitroso group is mechanistically similar to the addition to aldehyde. The electron pair of the nucleophile molecule attacks on the nitrogen atom of the nitroso group. In such a way a high structural diversity of products can be formed from the same nitroso compound only by selecting different nucleophiles. On the other hand the nitroso molecule itself is a nucleophile but with two nucleophilic centers, N and O atom. This property gives nitroso compounds a special role in the designing of organic synthesis of heterocyclic compounds. Recently, a series of new synthetic methods were developed in which nitrosoaromatic molecules appear as mediators in preparation of natural product building blocks with high enantiomeric purity.

The fact that nitroso function possesses two heteroatoms with n-electrons, and also π-electrons in the nitrogen-oxygen double bond, can be used for the synthesis of a wide variety of metal complexes. More than ten structural types of complexes are known in which the nitroso group behaves as a donor of the lone electron pairs situated on the N and O atoms, or as π donating ligands. This property is fundamental for the behavior of nitrosoaromatic compounds in biological systems, especially for their ligation with metals in porphyrin consisting biomolecules.

As a rule, nitrosoaromatic compounds appear in biological systems as intermediates, mostly on the pathway of either oxidation of amines or reduction of nitro compounds. Toxicity of nitro compounds is basically explained by their partial reduction to nitroso derivatives, which in turn can cause DNA damage. As good targets for the nucleophilic addition, nitrosoaromatic molecules react in biological systems with biomolecules that have sulfur atoms as potential nucleophiles. A series of reactions with substances such as glutathion, or proteins that have cysteine residues, are typical for the nitroso compounds. Only recently, the biosynthesis of the natural aromatic nitroso derivative was discovered, which is the final product rather than the reaction pathway intermediate.

Electronic structure of NO group with high energy HOMO orbital and possibility of $n <- \pi^*$ and $\pi <- \pi^*$ transitions is responsible for pronounced photochemical and electrochemical reactivity. Nitroso molecules readily form reactive nitroxide radicals. Because of their relative stability, aromatic nitroxides serve as spin traps for detections of other radicals that appear in photo- or electrochemical systems.

Multinuclear NMR spectra also reveals special characteristics not observed for most of the organic molecules. For instance, ^{15}N chemical shift anisotropies belong to the largest reported in the literature. The ^{17}O nucleus NMR spectroscopy is under development because of the improvements of the resolution of signals obtained by instruments available today. According to recent literature reports, nitrosoarenes can be used as a convenient test and model system for developing the ^{17}O NMR spectroscopy more generally. Significant progress have been made in the last decade in the field of dynamic NMR, anisotropy of chemical shifts, and anisotropy of quadrupole resonance just on the basis of detailed spectroscopy studies of aromatic nitroso molecules.

Undoubtedly, aromatic nitroso compounds are fascinating substances with extraordinary properties that comprise problems and concepts not only for this class of compounds, or for organic chemistry, but also for chemistry in general.

This book is divided in five chapters. The first chapter is focused on the structural properties of the aromatic nitroso molecules, mostly substituted nitrosobenzenes. The geometries of the characteristic examples calculated by high-level quantum chemical methods are compared with the structures obtained by X-ray diffraction analyses. The differences between the monomer and dimer forms of molecules are commented. An overview of the basic preparative methods is represented in the second chapter. Since most of the methods were already reviewed in few articles by Gowenlock and collaborators, the interest in this work is oriented more to the recent literature reports, as well as to some of our newly developed methods of solvent-free preparations. Physical organic chemistry and spectroscopy are discussed in the third, and the largest, chapter. Dimerization equilibria, structure-reactivity relationships, electronic and steric influences of functional groups on the molecular properties, electrochemistry, photochemistry, and reactivity are discussed in details. Typical reactions of nitrosoaromatic compounds and the corresponding characteristic reaction mechanisms are systematized and commented. The theme of the fourth chapter is organometallic chemistry that is of special interest because of a diversity of structures of complexes, which were discovered. The last, fifth, chapter deals with the role and behavior of aromatic molecules in the biological systems.

Although with the same functional group, there is difference between aromatic and aliphatic nitroso compounds, not only in their basic structural characteristics, but also in their use and importance in synthetic chemistry, mechanistic and physical-organic chemistry, and, especially in their biologically interesting activities. Although both are known as emitters of physiologically important NO molecules, nitrosobenzene derivatives are recently investigated more in biological systems because of their role as intermediates in the biodegradation of benzene-based pollutants.

Synthesis of aliphatic nitroso compounds is in principle simpler, most of them can be prepared by the addition of nitrosyl-chloride to the carbon-carbon double bond. Preparation of the aromatic nitroso compounds required development of a series of methods, none of them particularly efficient. As we will see later in this book, the most evident characteristic of the chemistry of aliphatic compounds is their facile rearrangement to the corresponding oximes by hydrogen shift from the α-carbon atom.

Possibility of derivatization of the aromatic ring makes aromatic nitroso molecules useful as appropriate models for studying a wide variety of chemical phenomena, starting from the reactivity-selectivity studies to the solid-state phenomena and processes on the surfaces. Replacement of functional groups on the same structural moiety of the aromatic ring opens opportunities to obtain a *ceteris paribus* condition for studying specific phenomena in physical organic chemistry and spectroscopy.

Chapter 1
Molecular Structures

Basic characteristics of the molecular structures of the most common aromatic C-nitroso molecules are described. Geometries and atomic charges of the molecules obtained by high-level quantum-chemical calculations are compared with the structural parameters provided from the crystallographic X-ray diffraction methods.

1.1 Calculated Molecular Structures

Almost all C-nitroso compounds exist in two forms, bluish or green colored nitroso monomers and colorless azodioxide dimers. This interplay between two forms represents perhaps the most interesting property of this class of organic compounds. Tendency of compounds to appear in one or in another form depends on the structural factors but even more on the conditions of the immediate environment, solution, crystal phase, supramolecular self-organization, and, of course, on the thermal conditions. Since the problem of dimerization is a matter of special chapter in this book, here we wish on the first place to discuss molecular structure in more details. Parent aromatic nitroso compound is nitrosobenzene, which will be used as a referent structure. Substitution on the benzene ring can basically have two kinds of influences on chemical behavior. Steric effect, which affects orientation of the NO group relative to the plane of the benzene ring, is present in the case of *ortho* substituted nitrosobenzenes. Electronic effect on the chemical reactivity of the nitroso group depends strongly on the nature of the *m*- and *p*-substituents. While electron donors, such as methoxy group, deactivate NO function, the electron-withdrawing substituents enhance its reactivity, especially towards dimerization. Influence of electron donors could basically be explained by using the resonance structures. As it is shown in the Scheme 1.1, the C–N bond becomes stronger and shorter, because of its partially double bond character.

The extent of such electron donation can be rationalized by details of molecular structures. Geometries as well as the Mulliken atomic charges of differently substi-

H. Vančik, *Aromatic C-nitroso Compounds*, DOI 10.1007/978-94-007-6337-1_1,

Scheme 1.1

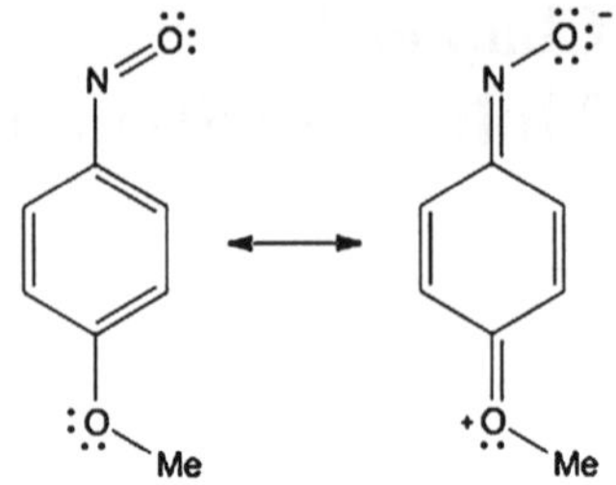

tuted nitrosobenzene molecules (shown in italics in the Fig. 1.1) were calculated by using high level quantum chemical methods (B3LYP/6-31G**) [1].

The calculated energies for the structures from the Fig. 1.1 are represented in Table 1.1.

Changes in geometry as well as in atomic charges in going form monomers to dimers can be recognized by comparison with the calculated structures of *Z*- and *E*-azodioxides (Fig. 1.2).

The energies for the differently substituted nitrosoaromatic dimers are represented in the Tables 1.2 and 1.3.

In the presented examples, the geometry parameters that are of interest for the analysis of chemical behavior of nitroso molecules, the NO, CN, and NN bond lengths are relatively insensitive to the substitution on the benzene ring. In the parent nitrosobenzene molecule, these bond lengths are calculated to be 1.223, and 1.441 Å, respectively. By using the AIM (Atoms in Molecules) method, Ejsmont [2] modified the NO distance to 1.214 Å. Interatomic NO distances fall between 1.223 and 1.224 Å. However, there are two exceptions, *p*-methoxy-, and *p*-nitronitrosobenzene. While the former molecule has longer bond length, the NO distance of the nitro derivative is shorter. Similarly, the average CN bond is between 1.436 and 1.441 Å long, but in *p*-methoxy-, and *p*-nitronitrosobenzene molecules the values are 1.428, and 1.450 Å, respectively. As we will see in following chapters of this book these two derivatives afford special behavior. While the methoxynitrosobenzene dimerize rarely and only under special conditions, nitronitrosobenzene readily forms azodioxides with a wide variety of nitroso monomer partners.

In the nitroso dimers, the NO bonds are longer (1.256–1.259 Å in *Z*-dimers, and 1.268–1.274 Å in *E*-dimers) because of the single bond character. In contrast to the trends in monomers, deviation from the average values calculated for the *p*-OCH_3 and *p*-NO_2 are smaller. Evidently, the NO bonds in dimers are less sensitive on the character of the substituent. The CN bonds in dimers are only slightly longer than in corresponding monomers, and appear in the narrow range of values (1.444–1.447 Å) for *Z*-dimers. In *E*-dimers, the *p*-methoxy derivative with its CN bondlength of 1.447 Å deviates from the average value of 1.450–1.453 Å. The bond between N-atoms (1.332–1.335 Å) is practically insensitive to the change of the *p*-substituent.

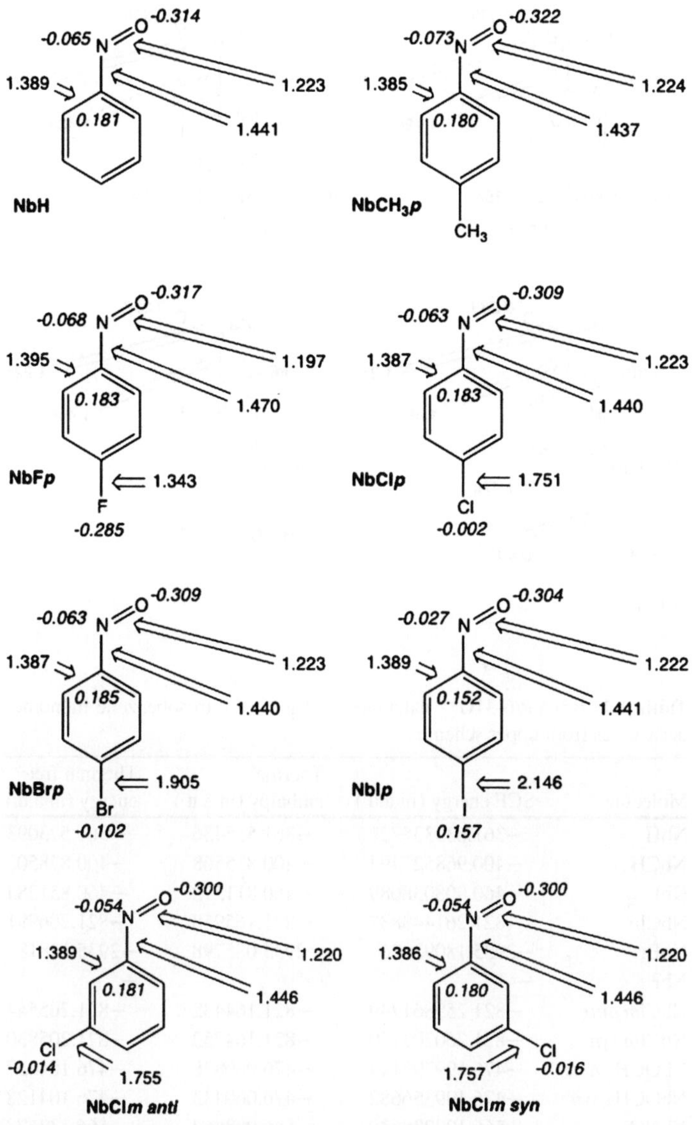

Fig. 1.1 B3LYP/6-31G** calculated structures of substituted nitrosobenzene molecules. Mulliken atomic charges are shown in *italics*. Bond lengths are in Å

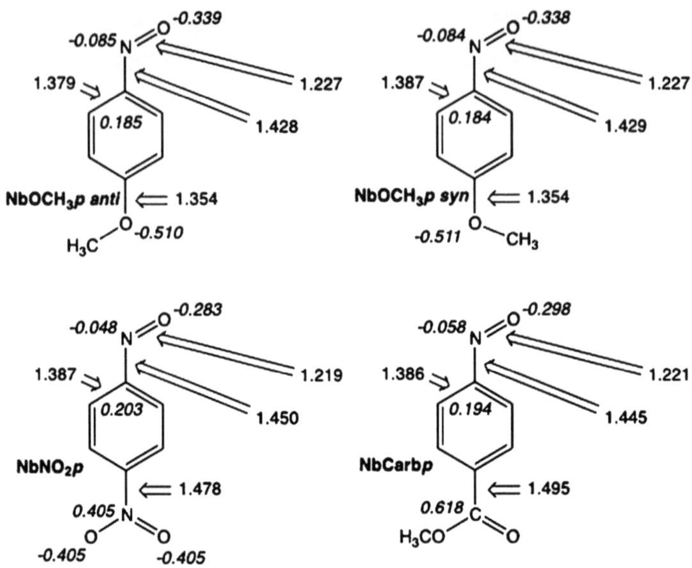

Fig. 1.1 (continued)

Table 1.1 B3LYP/6-31G** calculated energies of nitrosobenzene monomer derivatives from upper scheme

Molecule	SCF energy (in a.u.)	Thermal enthalpy (in a.u.)	Thermal free energy (in a.u.)
NbH	−361.639335721	−361.535436	−361.573093
$NbCH_3$	−400.968527494	−400.835568	−400.878503
NbF	−460.908038089	−460.811572	−460.851284
NbCl*p*	−821.261448837	−821.165958	−821.206991
NbBr	−2935.18093926	−2935.085798	−2935.12818
NbI	–	–	–
NbCl*m anti*	−821.259861749	−821.164432	−821.205544
NbCl*m syn*	−821.260203770	−821.164752	−821.205830
$NbOCH_3$ *anti*	−476.199793144	−476.060838	−476.104437
$NbOCH_3$ *syn*	−476.199356682	−476.060443	−476.104128
$NbNO_2$	−566.197238630	−566.088667	−566.133677
NbCarb *syn*	−589.583959752	−589.432928	−589.432928
NbCarb *anti*	−589.583950939	−589.432902	−589.482113

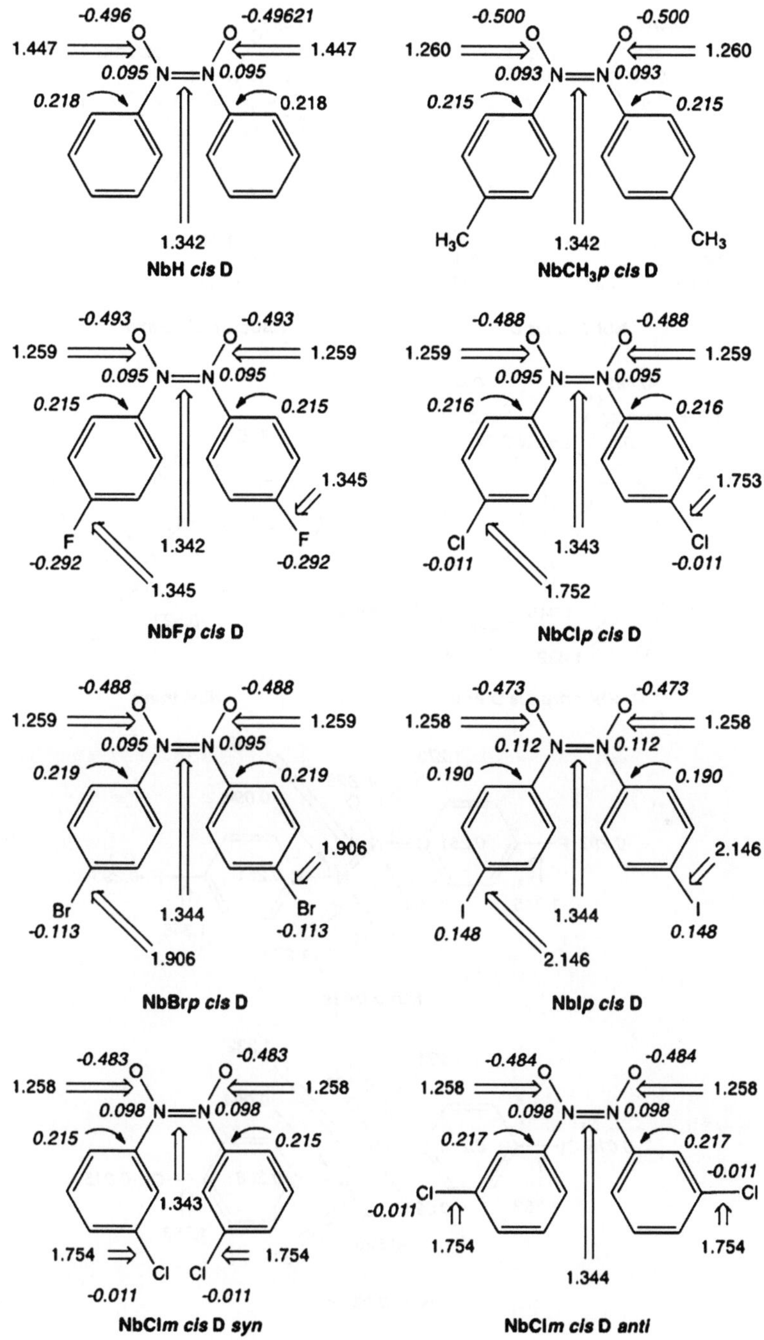

Fig. 1.2 B3LYP/6-31G** calculated structures of nitrosobenzene dimers. Mulliken atomic charges are shown in *italics*

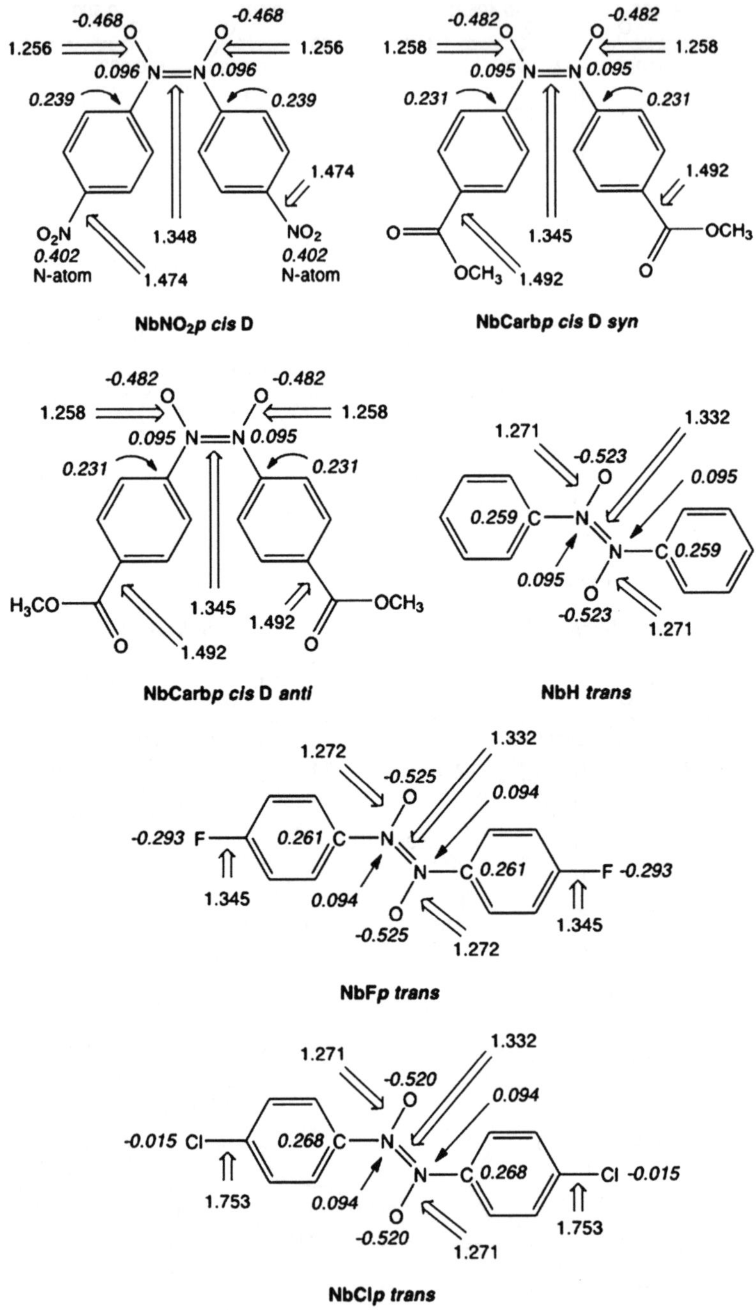

Fig. 1.2 (continued)

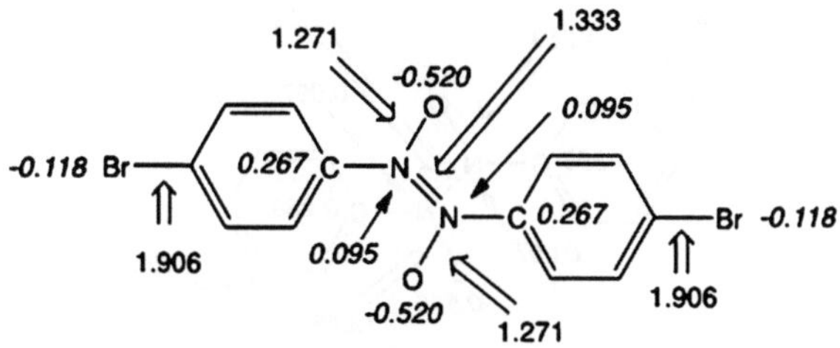

NbBrp *trans*

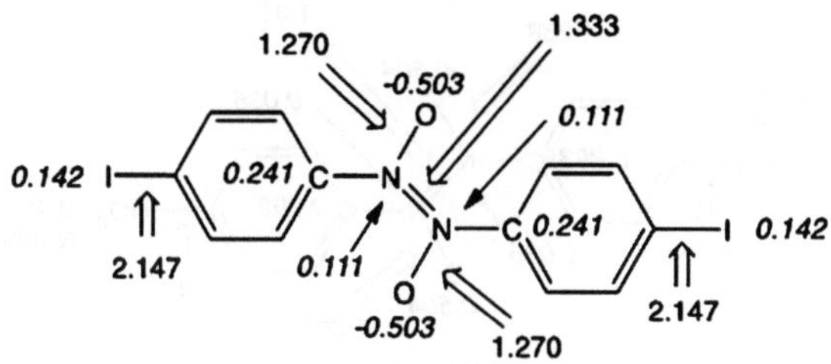

NbIp *trans*

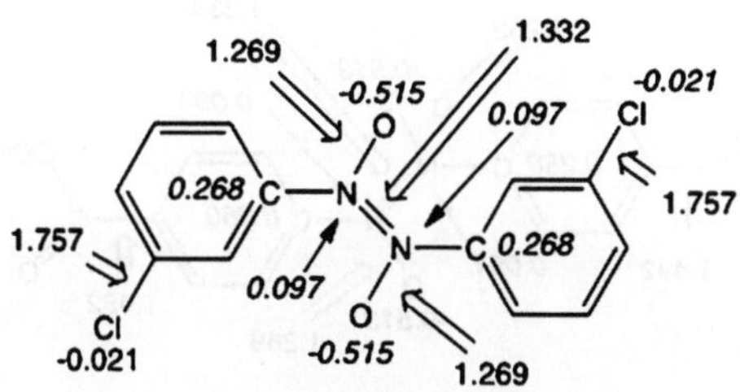

NbCl*m trans* syn

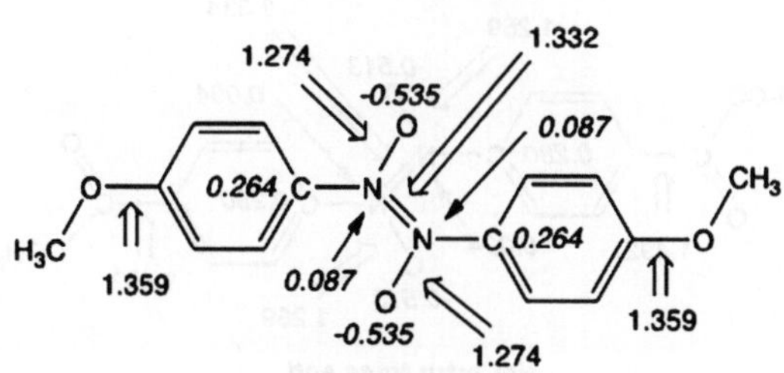

NbOCH$_3$*p trans*

Fig. 1.2 (continued)

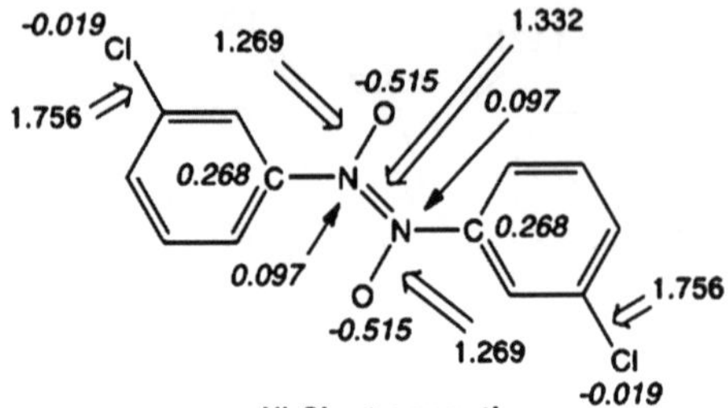

NbClm *trans anti*

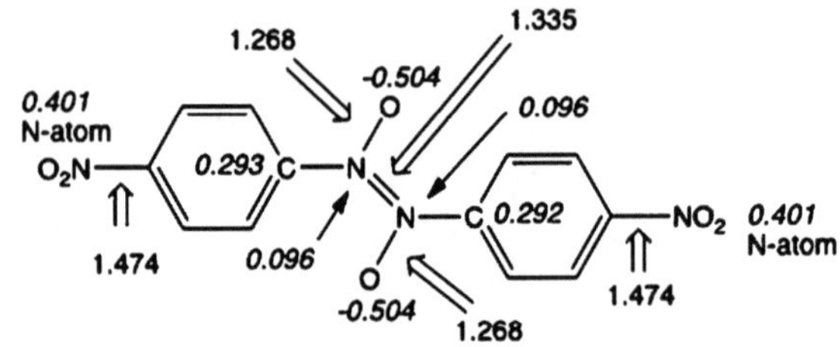

$NbNO_2p$ *trans*

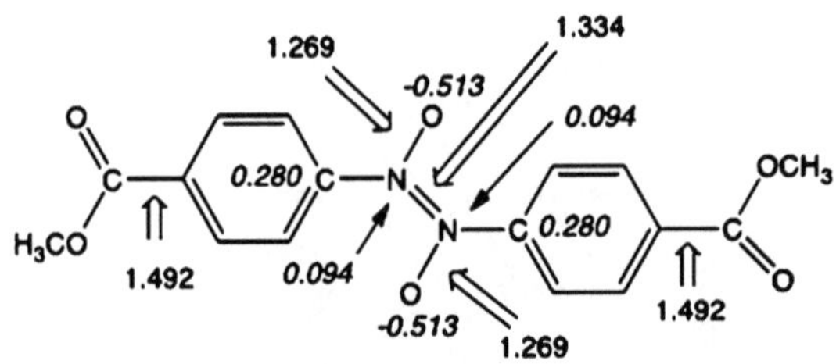

NbCarbp *trans syn*

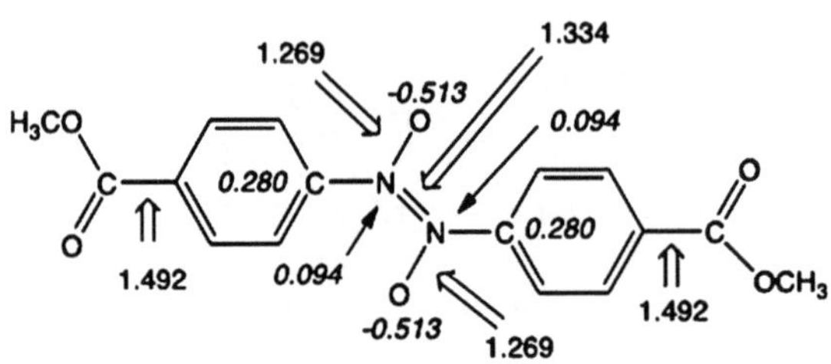

NbCarbp *trans anti*

Fig. 1.2 (continued)

Table 1.2 B3LYP/6-31G** calculated energies of the *E*-nitrosobenzene dimers for the derivatives represented in the Scheme 1.2

Molecule	SCF energy (in a.u.)	Thermal enthalpy (in a.u.)	Thermal free energy (in a.u.)
NbH trans D	−723.285248415	−723.073120	−723.127662
$NbCH_3$ trans D	−801.942807100	−801.672633	−801.736168
NbF trans D	−921.821666364	−921.624382	−921.683036
NbCl trans D	−1642.52999882	−1642.334684	−1642.395667
NbBr trans D	−5870.36853798	−5870.173938	−5870.238023
NbCl m trans D syn	−1642.52747168	−1642.332298	−1642.392737
NbCl m trans D anti	−1642.52767477	−1642.332472	−1642.392875
$NbNO_2$ trans D	−1132.40445386	−1132.182811	−1132.251585
NbCarb m trans D syn	−1179.17734433	−1178.871038	−1178.947537
NbCarb m trans D anti	−1179.17727828	−1178.870956	−1178.947371

Table 1.3 B3LYP/6-31G** calculated energies of the *E*-nitrosobenzene dimers

Molecule	SCF energy (in a.u.)	Thermal enthalpy (in a.u.)	Thermal free energy (in a.u.)
NbH cis D	−723.096422232	−722.883275	−722.937050
$NbCH_3$ cis D	−801.738725001	−801.467067	−801.531362
NbF cis D	−921.557675731	−921.359402	−921.417428
NbCl cis D	−1642.28189680	−1642.085714	−1642.146319
NbBr cis D	−5865.29960750	−5865.103828	−5865.166989
NbI cis D	−744.693308521	−744.498172	−744.563508
NbCl m cis D syn	−1642.28057743	−1642.084445	1642.145113
NbCl m cis D anti	−1642.28049686	−1642.084398	−1642.145040
$NbNO_2$ cis D	−1132.08772690	−1131.864716	−1131.931944
NbCarb m cis D syn	−1178.85844584	−1178.550517	−1178.626428
NbCarb m cis D anti	−1178.85866819	−1178.550714	−1178.626582

1.2 Experimental Geometries

Experimental geometries measured for the single crystals by X-ray diffraction methods in principle correspond to the calculated values, but show larger variations in the case of differently substituted nitrosobenzenes [5–19]. The differences could be attributed to the influence of crystal packing, where the structure adopts the rigid geometry. The selected molecular geometry parameters are represented in Fig. 1.3. In monomers, the NO bond lengths fall between 1.200 and 1.228 Å. Generally, in nitroso compounds, this value is in the 1.13–1.29 Å range [3]. In dimers, the NO bond distances are the same as calculated and fall in narrow range between 1.25 and 1.28 Å.

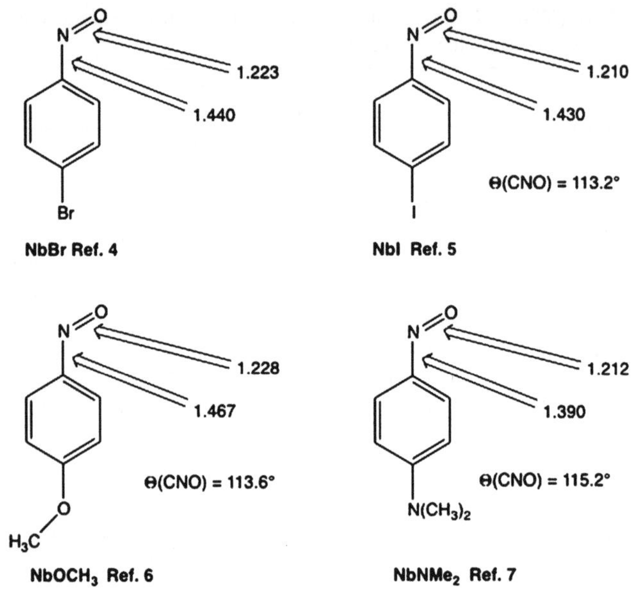

Fig. 1.3 Experimental geometries of nitrosobenzene monomers (X-ray diffraction)

The CN bond of monomer shows larger variability, and dependency of the *p*-substituent. The observed geometrical diversity could be explained by the resonance structure model, in which the electron donors such as methoxy or amino groups enhance the double bond character.

Very important structural property of azodioxides is the angle between benzene ring planes and the ONNO plane. The two planes are never coplanar except for the *Z*-*p*-methoxy dimer where this angle is close to zero value (**$NbOCH_3$ *E***, in the Fig. 1.4). If the benzene rings are substituted in the *ortho* position, the two planes are almost perpendicular to each other (structures **Nbfen *E***, **$NbMeONO_2$**, and **NbiPr *E*** in the Fig. 1.4).

As we can see later in this book, packing of the molecules in crystal has one of most important influence on the properties, especially chemical reactivity and phase transformations of the nitrosobenzene dimers. Most of the nitroso dimers are arranged in the crystal with the parallel planes of the benzene rings. Illustrative example is crystal packing of *p*-bromonitrosobenzene dimer, shown in the Fig. 1.5. The hydrogen-bonding network is in the Figure labeled by yellow lines. Oxygen atoms of the azodioxide group are 2.433 Å distant from the *m*-hydrogens of the benzene ring of the neighboring molecule. Bromine-bromine close contacts within the crystal plane are 3.68 Å long, and in the Fig. 1.5, they are labeled by red lines [4].

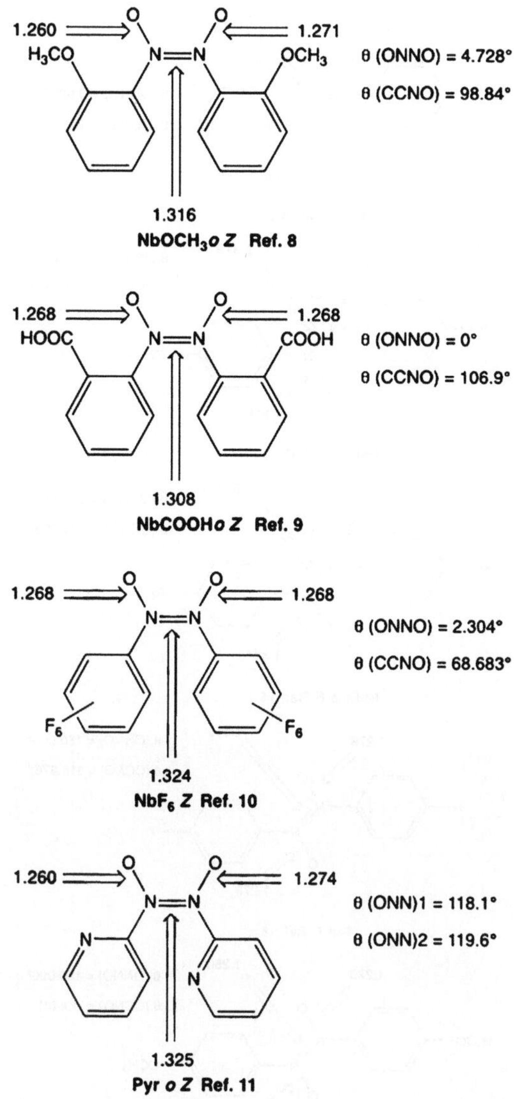

Fig. 1.4 Experimental geometries of nitrosobenzene dimers (azodioxides) obtained by X-ray diffraction

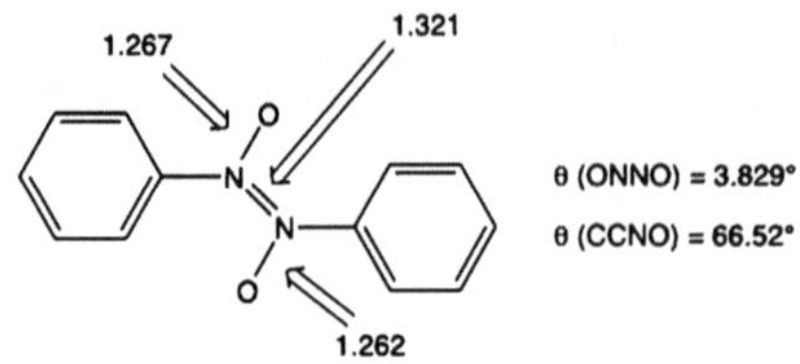

NbH *E* Ref. 12

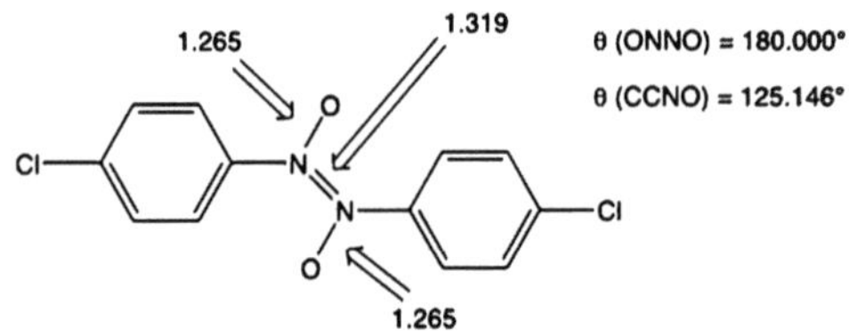

NbCl *p E* Ref. 13

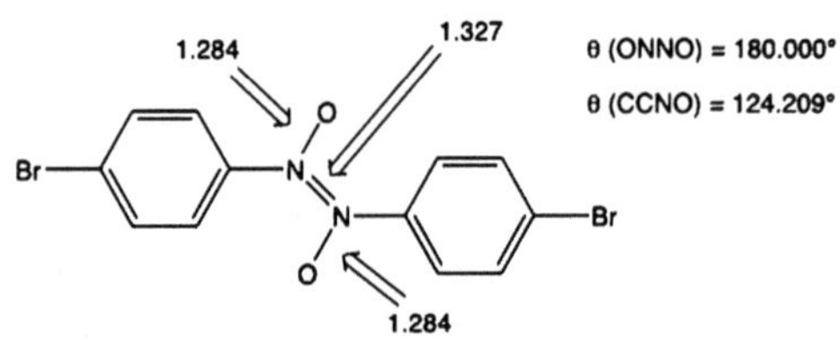

NbBr *p E* Ref. 14

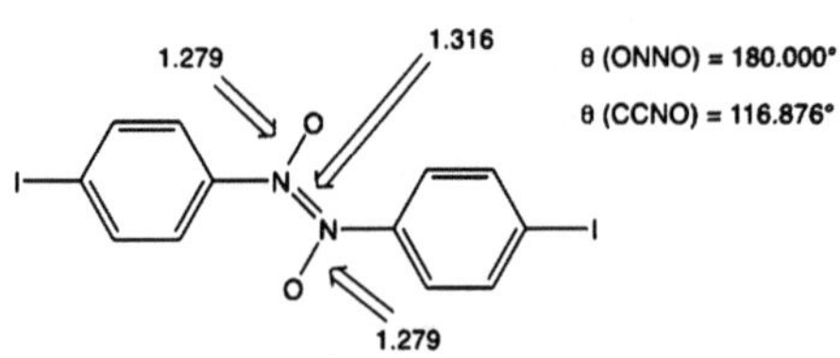

NbI *E* Ref. 16

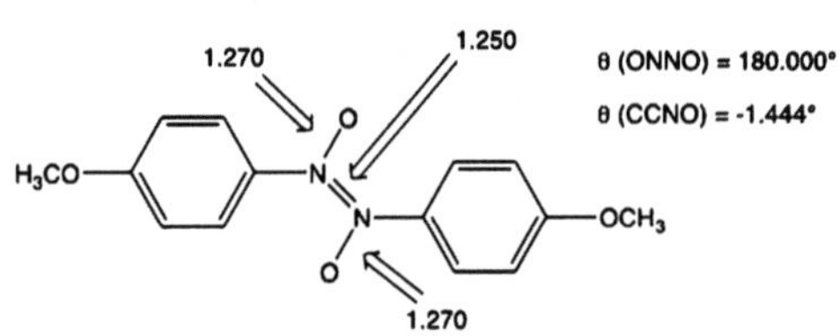

NbOCH$_3$ *E* Ref. 15

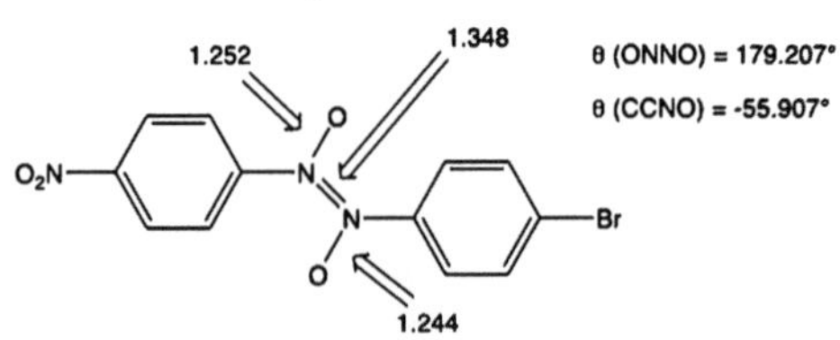

NbBrNO$_2$ *E* Ref. 14

Fig. 1.4 (continued)

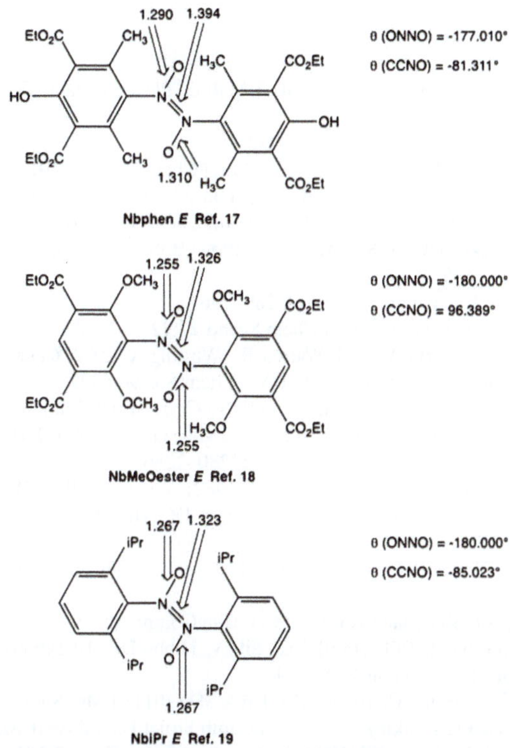

Fig. 1.4 (continued)

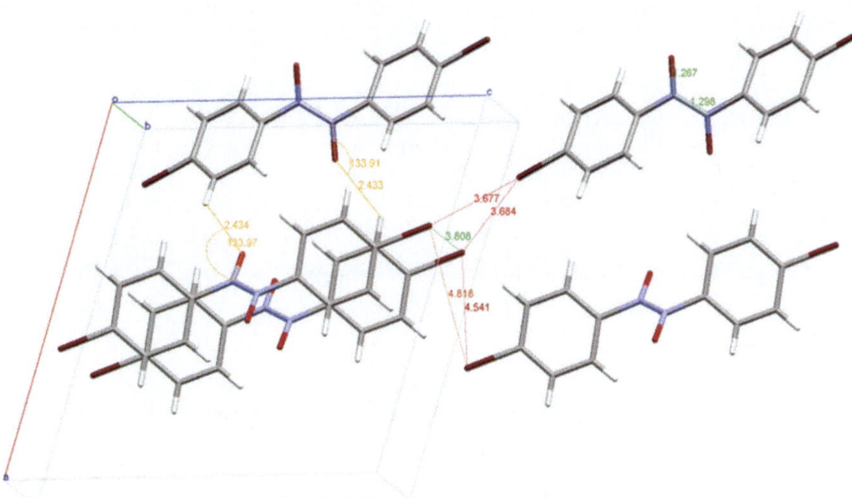

Fig. 1.5 Crystal packing of *p*-bromonitrosobenzene dimer. *Red lines* and the *green line* represent the bromine-bromine close contacts (Reproduced by permission from Ref. [4])

References

1. Biljan I, Cvjetojević G, Smrečki V, Novak P, Mali G, Plavec J, Babić D, Mihalić Z, Vančik H (2010) J Mol Struct 979:22–26
2. Ejsmont K (2007) Polish J Chem 81:985–992
3. Lee J, Chen L, West AH, Richter-Addo GB (2002) Chem Rev 102:1019–1065
4. Vančik H, Šimunić-Mežnarić V, Meštrović E, Halasz I (2004) J Org Chem 69:4829–4834
5. (a) Dieterich DA, Paul IC, Curtin DI (1974) J Am Chem Soc 96:6372; (b) Miao FM, Chantry D, Harper T, Hodgkin DC (1982) Acta Crystallogr B38:3152–3155; (c) Webster MS (1956) J Chem Soc 2841–2845
6. Talberg HJ (1979) Acta Chem Scand 33A:289–296
7. Romming C, Talberg HJ (1973) Acta Chem Scand 27:2246–2248
8. Lightfoot AP, Pritchard RG, Wan H, Warren JE, Whiting A (2002) Chem Commun 2072
9. Dieterich DA, Paul IC, Curtin DI (1974) J Am Chem Soc 96:6372
10. Prout CK, Coda A, Forder RA, Kamenar B (1974) Cryst Struct Commun 3:39
11. Gowenlock BG, Maidment MJ, Orrell KO, Šik V, Mele G, Vasapollo G, Hursthouse MB, Abdul Malik KM (2000) J Chem Soc Perkin Trans 2:2280–2286
12. Miao FM, Chantry D, Harper T, Hodgkin DC (1982) Acta Crystallogr B38:3152–3155
13. Fletcher DA, Gowenlock BG, Orrell KG, Apperley DC, Hursthouse MB, Malik KMA (1999) J Chem Res 202:1115
14. Halasz I, Biljan I, Novak P, Meštrović E, Plavec J, Mali G, Smrečki V, Vančik H (2009) J Mol Struct 918:19–25
15. Janbon S, Davey RJ, Shankland K (2008) CrystEngComm 10:279
16. Fletcher DA, Gowenlock BG, Orrell KG, Šik V, Hibbs DE, Hursthouse MB, Malik KMA (1996) J Chem Soc Perkin Trans 2:191–197
17. Alemasov YA, Slaschinin DG, Tovbis MS, Kirik SD (2011) J Mol Struct 985:184
18. Barnes JC, Chudek JA, Weakley TJR (2005) Camb Struct Data Base (CSD)
19. Gowenlock BG, McCullough KJ (1989) J Chem Soc Perkin Trans 2:551

Chapter 2
An Overview of Synthetic Methods for Preparation of Nitrosoaromatic Compounds

Outline of the basic preparative methods that include three possible ways of the synthesis of aromatic C-nitroso compounds are represented. Besides the classical methods of reduction of nitro- or oxidation of amino precursors, the approaches based on the direct nitrosation, solid-state syntheses as well as enzyme-catalyzed reactions are commented in more details. Since the nitroso compounds easily undergo formation of dimers and azoxy products, discussion about the preparation of these derivatives is added.

Nitrosobenzene as the parent aromatic nitroso compound has been for the first time prepared by Adolf Baeyer [1] who applied the reaction of diphenylmercury with NOBr. By the same type of reaction Baeyer also synthesized nitrosonaphthalene (Scheme 2.1).

As we have mentioned in the Introduction, since nitroso compounds easily dimerize to azodioxides, some of the synthetic methods, which were designed to prepare azodioxides also serve as methods for preparation of nitroso compounds. Because nitroso compounds appear as intermediates on the redox scale between limiting amino and nitro sides, both the synthetic options for their preparation are open, oxidation of amino-, or reduction of nitro-derivatives. Aliphatic derivatives can also be obtained by using other type of reactions, such as additions of nitrosylchloride on the carbon-carbon double bond, or different photoreactions. Additional complication with aliphatic nitroso compounds is their tendency to rearrange to oximes by [1,3] hydrogen shift from the alpha carbon atom (Scheme 2.2).

On the other hand, in the aromatic nitroso compounds, equilibrium between the nitroso and oxime form has been observed only in the *o*-hydroxy-substituted nitrosoaromatic molecules such as 1-nitroso-2-naphthol [2]. However, this rearrangement does not interfere with the preparation of the compound (Scheme 2.3).

Heteroaromatic nitroso compounds (Scheme 2.4) can also exist in such tautomeric forms as it is exampled by 4-nitroso-5-pyrazolone [3–5].

Application of redox reactions has two main difficulties. First is in finding proper conditions under which the reaction will not continue to the final oxidation or

H. Vančik, *Aromatic C-nitroso Compounds*, DOI 10.1007/978-94-007-6337-1_2,

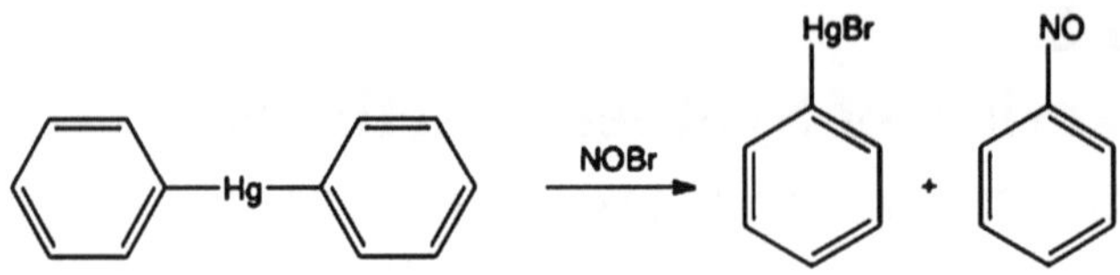

Scheme 2.1

Scheme 2.2

Scheme 2.3

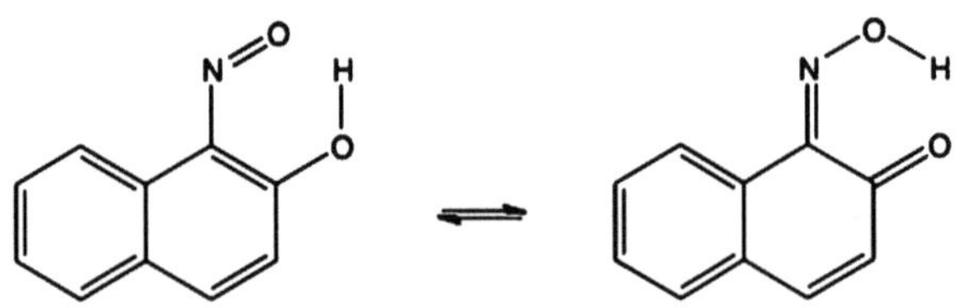

Scheme 2.4

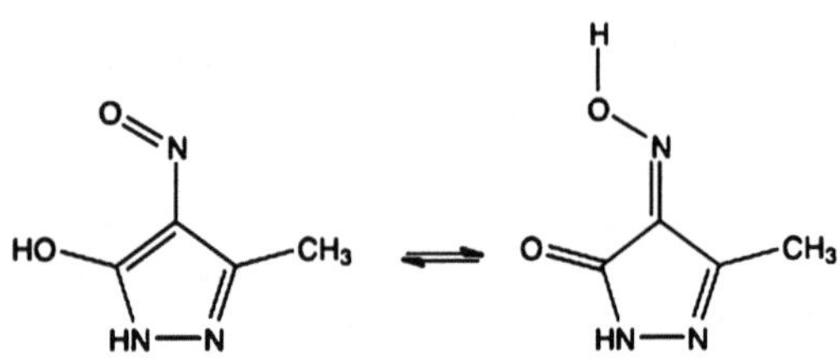

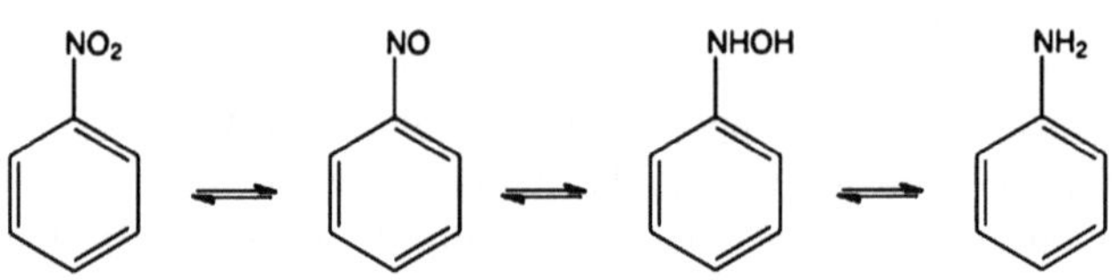

Scheme 2.5

reduction product, but instead, will stop when the nitroso intermediate is formed (Scheme 2.5). The second problem has its origin in the tendency of intermediates to form azoxides as byproducts. To avoid these difficulties, besides the optimized classical synthetic pathways, a series of specific reagents as well as methods were developed. Most recently, the solvent-free solid-state preparative methods combined with purification by sublimation were used successfully. The method will be discussed later in this Chapter.

2.1 Reductive Methods

Starting material for preparation of nitroso molecules by reductive reactions are nitro-derivatives, i.e. in the case of aromatic compounds they are substituted nitrobenzenes. Historically, there is more than a 100 years long tradition in developing the reductive methods to prepare nitroso- from nitrobenzene derivatives, especially for possible applications in industrial synthesis [6]. As reducing reagents and/or catalysts, metals, metal amalgams, and even metal oxides were used [7–9]. It has been shown from results of calculations on DFT levels of theory, there are specific mechanisms in which metal or metal oxide reductions include single electron transfer [10]. Some of manganese oxide catalysts became already classics in the large-scale preparations. While strong conventional reductive reagents, such as Sn, $SnCl_2 + HCl$, $Fe + HCl$, or H_2, Pt, transform nitro- to amino group, weaker reductive reagents, for instance Zn powder, can under controlled pH conditions yield hydroxylamine [11, 12], which in the next step can be re-oxidized to the nitrosobenzene by using $FeCl_3$ or $Na_2Cr_2O_7$ in sulfuric acid, as it is represented in the following equation. The method can widely be used in laboratory synthesis of nitrosoaromates. However, the yields obtained in these preparations are not always satisfactory because of the formation of byproducts, mainly azoxides and nitroaromates.

$$C_6H_5NO_2 \xrightarrow{\text{Zn, NH}_4\text{Cl}} C_6H_5NHOH \xrightarrow[\text{Na}_2\text{Cr}_2\text{O}_7]{\text{FeCl}_3\text{ or}} C_6H_5NO$$

In the second step, the oxidation of hydroxylamine, a variety of other mild oxidants such as periodic acid [13], periodates [14, 15] or silver carbonate [16] could be useful. Main disadvantage in using such an oxidative method is again the appearance of products of a side reaction. Namely, nitrosobenzene forms azoxybenzene byproduct in the condensation reaction with hydroxylamine intermediate [17].

A number of polycyclic aromatic nitro compounds, especially heterocyclic systems can be reduced to nitroso analogs also by addition of Grignard reagents followed by reaction with strong acid [18, 19]. Unfortunately such a method frequently yields a mixture of *ortho* and of *meta* substituted derivatives, as it is shown on the Scheme 2.6.

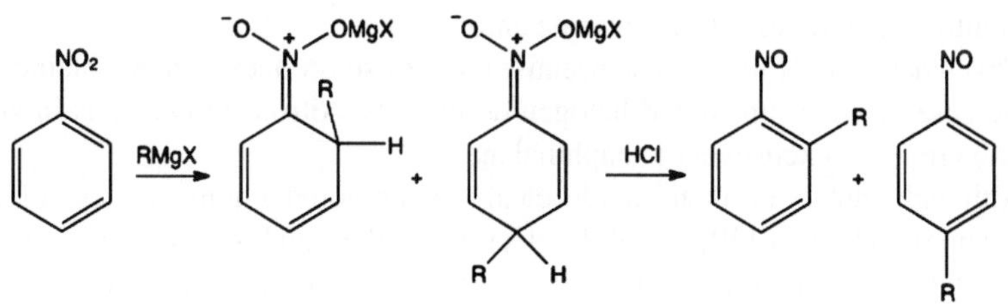

Scheme 2.6

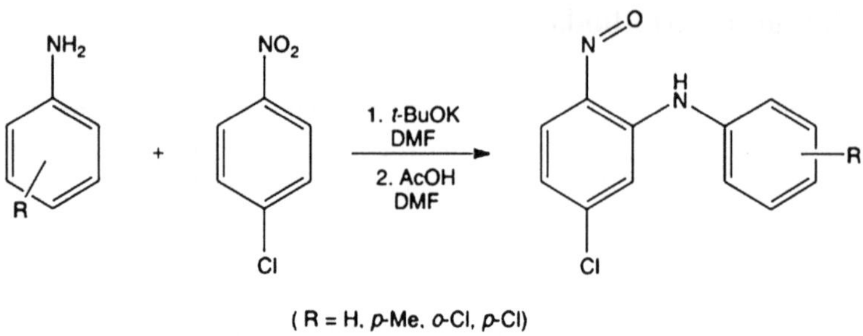

Scheme 2.7

Scheme 2.8

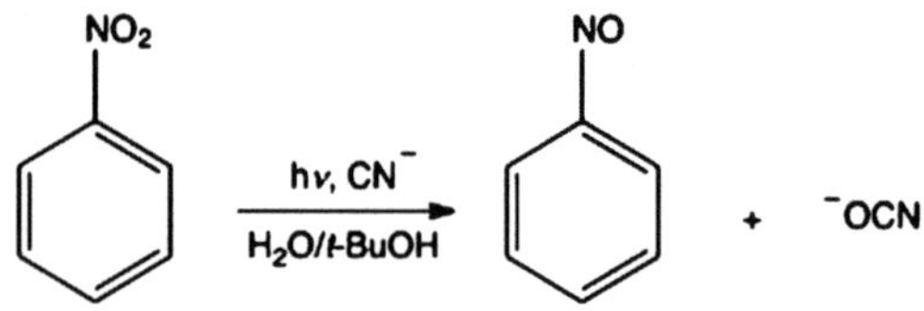

Substituted nitrosobenzenes can also be prepared by direct reaction of aniline with corresponding nitrobenzene. The 2-nitroso-*N*-arylanilines (Scheme 2.7) were synthesized in a one-pot reaction from anilines and 1-chloro-4-nitrobenzene [20, 21].

Beside such "classical" methods, a series of the electrochemical and photoreductions of nitroaromatic precursors were developed. In principle, electrochemical reductions follow the same reaction pathway through formation of hydroxylamine intermediate [22, 23]. These methods include either a standard electrochemical synthesis or specially constructed electrodes [24–27]. Most of differently substituted nitrobenzenes with substituents in all positions, *ortho*, *meta* or *para*, were reduced to nitroso derivatives with high yield. By using a gold electrode, the aromatic nitro compounds can be reduced to nitroso derivatives by electrochemical reaction on the electrode surface.

Photoreduction of substituted nitrobenzene in most cases does not end with formation of corresponding nitroso compounds. Rearranged products with substituted *ortho* positions are preferred. Of the synthetic value could be photoreaction of nitrobenzene in the presence of potassium cyanide (Scheme 2.8). Oxygen atom is from nitro group removed as a cyanate anion [28]:

This reaction is of limited use because the nitroso products can be obtained only in the case of *para* substituted halogen derivatives. Nitronaphthalene is in similar reaction reduced exclusively to naphthalene.

Although metal- or metal-oxide catalysed photoreductions are less effective than photooxidations [29], careful selection of the catalyst could be useful not only for the preparation of nitro-, but also for nitrosoaromatic compounds. One of the promising catalysts is titanium dioxide, which promotes photoreduction of nitrobenzene [30, 31].

Some amount of nitrosobenzene was also isolated as one of products of enzymatic reduction of nitro compounds. It was already known from the industrial experience that some flavoenzymes (called "Old Yellow Enzyme", OYE) are effective in biodegradation of nitro-explosives [32]. Recently, a wide variety of stereoselective reductions supported by enzymes such as OPR1 and OPR3 from *Lycopersicon esculentum* [33], OYE from *Saccharomyces carsbergensis* and OYE2-3 from *Saccharomyces cerevisiae* [34], NCR from *Zymomonas mobilis* [34], YqjM from *Bacillus subtillis* [33], PETN reductase from *Enterobacter cloacae* [35] were tested. Nitroreductase from *Salmonella typHimurium* [36] transforms nitrobenzene in the mixture consisting from nitrosobenzene, phenylhydrazine, as well as diphenylazodioxide.

2.2 Oxidative Methods

Preparation of aromatic nitroso compounds from amino precursors is more practical because of numerous available oxidative methods. A series of peroxoacids may be applied for oxidation of aromatic amines, peroxomonosulfuric acid (Caro's acid) [37, 38], peroxyacetic acid [39], 3-chloroperoxybenzoic acid [40], or peroxyformic acid [41, 42]. Hydrogen peroxide in combination with organometallic catalysts (peroxotungstophosphate) [43], $[Mo(O)(O_2)_2(H_2O)(hmpa)]$ [44], *cis*-$Mo(O)_2(acac)_2(acacH{=}MeC(O)Me)$ [45], methylrhenium trioxide [46], oxoperoxo(pyridine-2,6-dicarboxylato)(hmpa)molybdenum(VI) [47], as well as with inorganic salts such as sodium tungstate or sodium-EDTA [48] is known to give nitrosoaromates in good yields. The requirements for environmentally adequate methods, the green-chemistry preparations, prompted scientists to find new catalysts such as heteropolyacids (HPAs) [49]. In combinations with hydrogen peroxide HPAs can effectively oxidize aniline to nitrosobenzene [50]. Oxidation can be performed in triphase system comprising an aqueous solution of hydrogen peroxide, an organic solvent, and Aliquat 336 surfactant. It must be pointed out, that most of the investigated HPAs afford almost 100 % conversion with high selectivity in favor of nitrosobenzene over nitrobenzene [50].

Classical oxidants such as potassium permanganate in mixture of formaldehyde and sulfuric acid [51] could also be used. Perhaps the best results were obtained by using mixture of sodium peroxosulfate with sodium sulfate and sodium hydrogensulfate, the reagent known under popular name *Oxone* [52, 53].

2.3 Solid-State Syntheses

The last mentioned reagent (Oxone) is also very convenient for the recently investigated solvent-free reactions. Grinding of powdered samples of substituted anilines with oxone yields nitroso compounds, which can easily be sublimed out of the

Table 2.1 Yields of the solid-state preparation of differently substituted nitrosobenzenes (Data from Ref. [53])

Substituent	*p*-Cl	*p*-Br	*p*-I	*p*-NO_2	*p*-CH_3
Yield %	70	80	85	26	52
Purity %	80	95	92	90	69
Milling time/min	20	20	20	30	20
$NaHCO_3$ addition	Yes	Yes	Yes	No	Yes

reaction mixture [54]. The reaction is more efficient if sodium hydrogencarbonate is added as a component for neutralization. The results of oxidation of differently substituted nitrosobenzenes are represented in the Table 2.1.

The only impurity found after sublimation was corresponding nitrobenzene.

Reactions in porous medium represent a perspective methodology in organic functional group transformations. Pillared interlayered clays (PILCs) are class of such a porous material with cavities large enough to serve for depositing metal catalysts [55]. Titania pillared montmorillonite clays with TiO_2 nanosized particles as a catalytic complex afford very high selectivity in production of nitrosobenzene and/or azoxybenzene by oxidation of aniline with H_2O_2. While low concentration of the catalyst results with 80 % yield of nitrosobenzene, high concentration of catalyst promotes high yield, almost 99 %, of azoxybenzene [56].

2.4 Enzyme Catalyzed Oxidations

Oxidation of arylamines into their corresponding nitroso derivatives is also possible by enzyme catalysis. Recent use of crude chloroperoxidase isolated from *M. paradisiaca* for the *N*-oxidation of aromatic amines has demonstrated the efficiency of enzyme-assisted syntheses [57]. Since the enzyme active center is iron porphyrin, the proposed mechanism [57, 58] includes *N*-activation of amine by the trivalent iron cation radical complexed with oxygen. The mechanism is represented in Scheme 2.9.

In previous example, as well as in most aromatic nitrosations described to date, the nitroso derivative appears exclusively as an intermediate that is further oxidized to the corresponding nitro derivative. A series of cofactors are already known to be included in the enzymatic aminoxidations [59]. However, recent discovery of the natural enzymatic system opens a new perspective in the application of enzymes for the preparations of aromatic nitroso compounds. Noguchi et al. [60] have discovered the enzyme assigned as NspF from *Streptomyces murayamaensis* that catalyses biosynthesis of the nitroso natural product 4,3-HNBAm starting with the corresponding amine derivative. The active site of the enzyme includes two copper atoms which are able to bind oxygen atom and form the peroxo-dicopper(II) complex. Amino substrate reacts with oxygen from the complex and after elimination of water transforms directly to the nitroso derivative (Fig. 2.1).

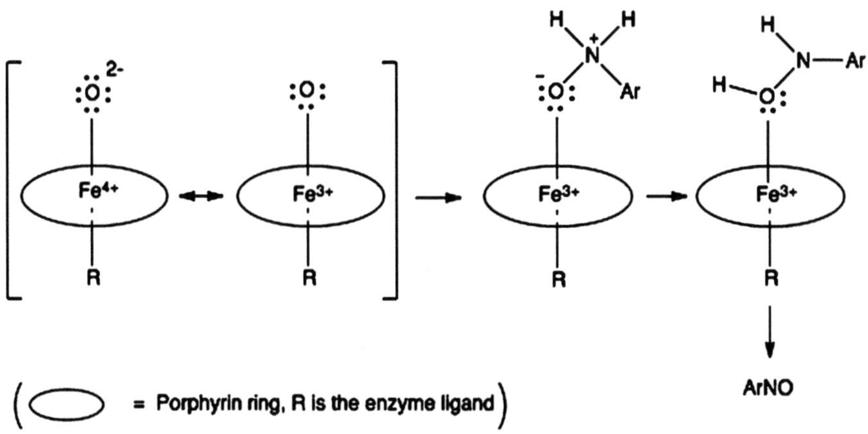

Scheme 2.9

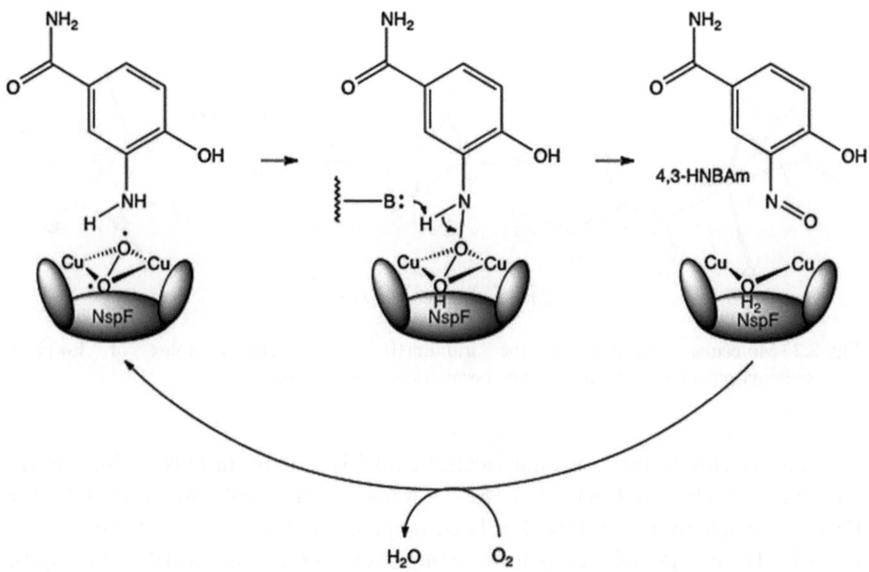

Fig. 2.1 Mechanism of the oxidation of the substituted aniline derivative by the enzyme NspF from *Streptomyces murayamaensis*

2.5 Direct Nitrosation

Relative stability of nitrosonium cation, NO^+, affords good opportunity for direct nitrosation of arenes by electrophylic substitution reactions (Scheme 2.10).

The available methods differ in source of nitrosonium ion. In most of methods nitrosonium ion is prepared *in situ*, starting with gaseous nitric oxide in combination

Scheme 2.10

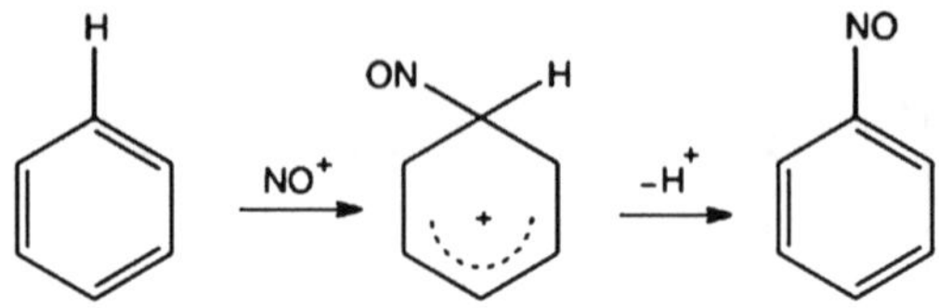

Scheme 2.11

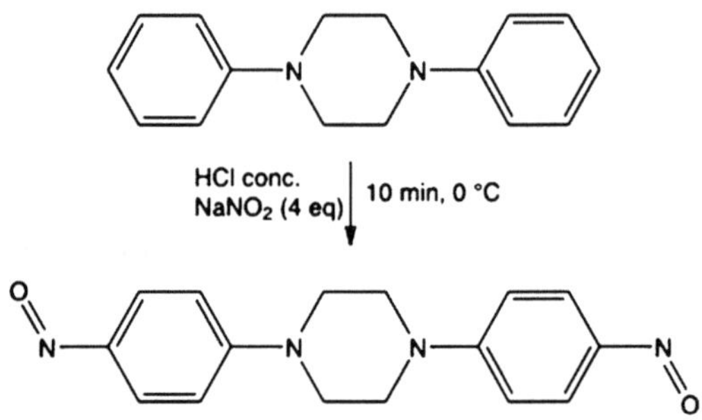

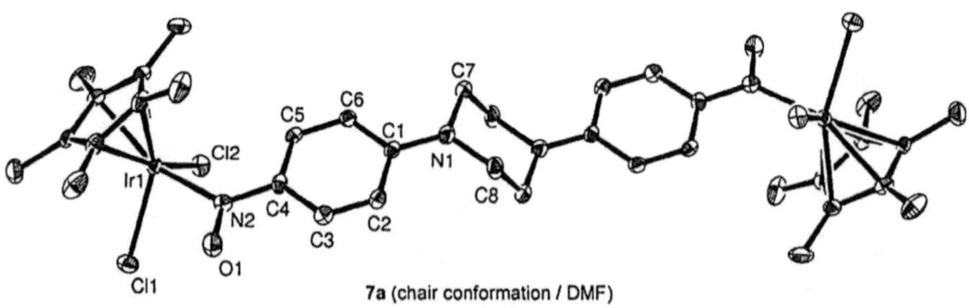

Fig. 2.2 Molecular structure of the iridium(III) metallocene complex of 1,4-bis(4-nitrosophenyl)piperazine (Reproduced by permission from the Ref. [63])

with strong (for instance trifluoroacetic) acid [61], or by mixing sodium nitrite with acid (acetic, hydrochloric) [62]. An interesting dinitroso derivative, 1,4-bis(4-nitrosophenyl)piperazine, has been prepared by this way (see Scheme 2.11) [63, 64]. The compound readily forms complexes with the iridium(III) metallocene, as in the example represented in the Fig. 2.2.

The method has also been successfully used in preparations of nitrogen containing heteroaromatic compounds. For *para*-nitrosation of *N*- or *N,N*-substituted anilines these methods are only partially useful, because in the case of *N*- monosubstituted anilines, the *N*-nitroso compounds as intermediate products were obtained.

Transformations of *N*-nitroso to *C*-nitrosoaromatic derivatives are known. An interesting example is thermal single electron oxidation or photoreaction of differently substituted *N*-nitroso-*N*-phenylhydroxylamine ammonium salt that yields azodioxides and/or corresponding nitrosobenzenes [65]. The reaction is also applicable for generation of NO (Scheme 2.12).

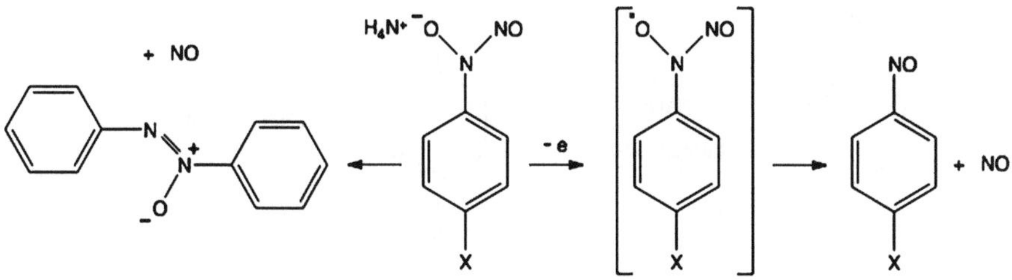

Scheme 2.12

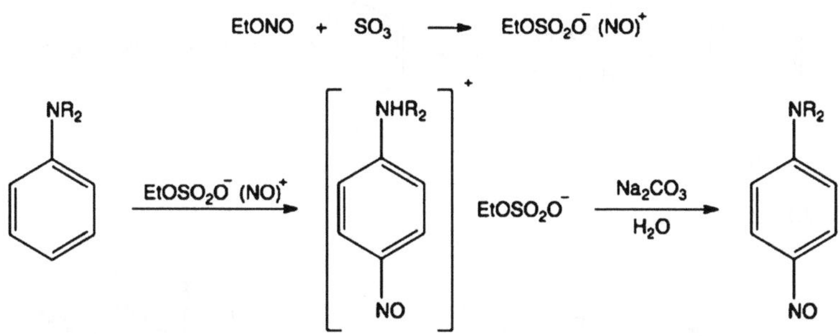

Scheme 2.13

Since nitrosonium ion can be persistent as a salt of strong Lewis acids, suitable reagent for nitrosation of aromatic compounds is nitrosonium tetrafluoroborate that is especially efficient for direct nitrosation of alkyl- and polyalkyl-substituted benzenes [66].

For nitrosation of arenes and arylmetal compounds the electrophilic nitrosonium agents have been systematically investigated [66–70]. However, the wider use of this type of reactions in the synthesis was limited because of the low yield, side products, and high toxicity of some organometal compounds. Efficient reagent for the preparation of *p*-dialkylamino nitrosobenzenes is nitrosonium ethylsulfate [67]. The reagent can be prepared *in situ* from EtONO and SO_3 (Scheme 2.13).

Treatment of the stable intermediate salt with sodium carbonate solution yields corresponding nitrosobenzene derivative.

Recently, Molander and Cavalcanti [71] have developed very efficient method for nitrosation of the aromatic compounds (Scheme 2.14) by the reaction of aryl-trifluoroborate salt with $NaNO_2$ or $NOBF_4$.

In the proposed mechanism (Scheme 2.15), the first step is attack of the nitrite ion on the boron atom, followed by the rearrangement in which the NO^+ group behaves as an electrophile.

The reaction is also applied for the nitrosation of heteroaromatics, and for *in situ* preparation of less stable nitrosoaromatic reagents, which can be used in syntheses

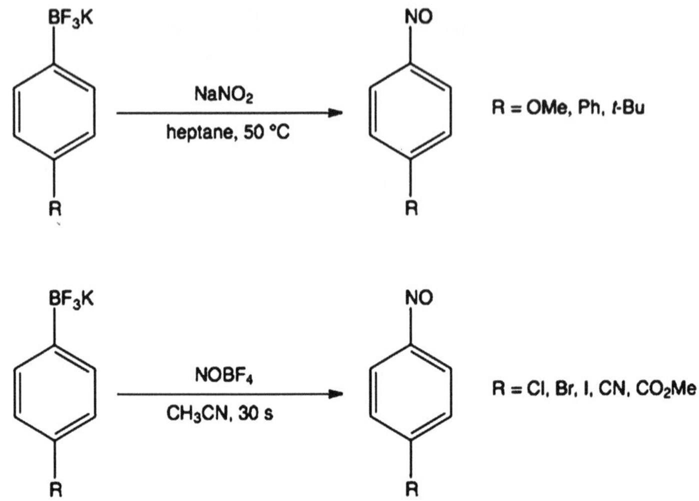

Scheme 2.14

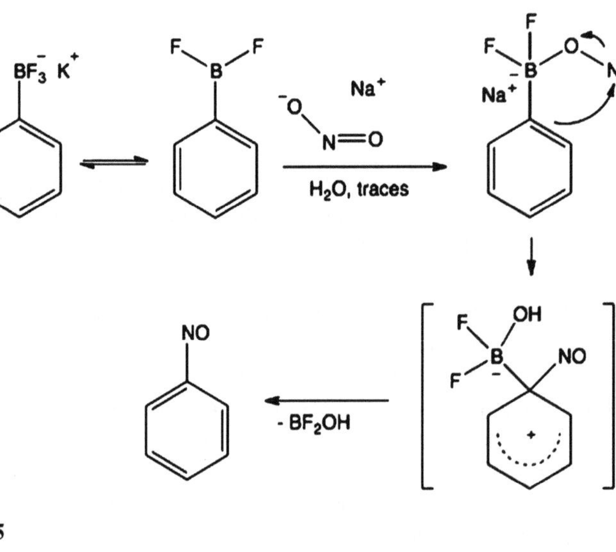

Scheme 2.15

of polycyclic structures. For instance, the one pot reaction of trifluoro(isoquinolin-5-yl)borate with $NOBF_4$ and the corresponding diene yields the Diels-Alder adduct (Scheme 2.16).

Direct nitrosation of the heteroaromatic derivative is possible if the starting compound possesses trimethylsilyl substituent, as in the case of 4-trimethylsilyl-pyrazol (Scheme 2.17). Although not in high yield, the reaction of the trimethylsilyl pyrazol with sodium nitrite and trifluoroacrtic acid gives corresponding nitroso derivative [72].

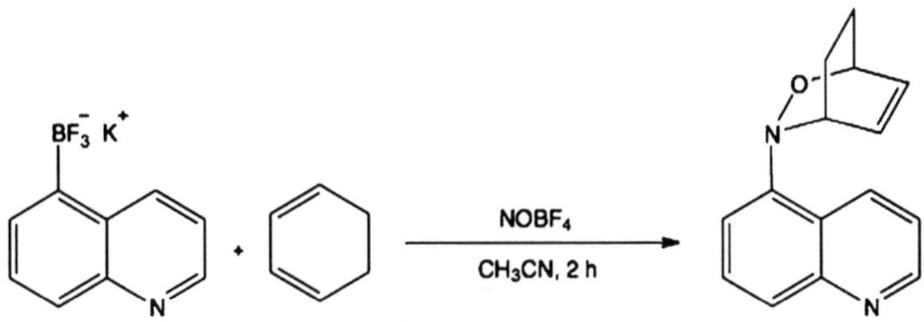

Scheme 2.16

Scheme 2.17

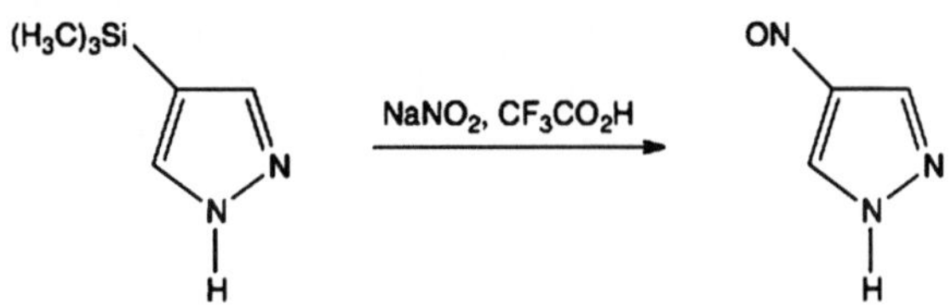

Similarly the aromatic trimethylstanyl compounds can be transformed to the corresponding nitroso derivatives by the reaction with NOCl [73]:

$$ArSnMe_3 + NOCl \rightarrow ArNO + Me_3SnCl$$

Methylated nitrosobenzene derivatives can also be prepared by 50–90 % yield by reaction of aromatic organothallium compounds with nitrosyl chloride generated *in situ* [74].

$$ArTlCl_2 + NOCl \rightarrow ArNO + TlCl_3$$

Direct nitrosation can be achieved by using the metal complexes as mediators. Insertion of NO^+ into the ruthenium-aryl bond of cyclometalated ruthenium(II) complex led to the formation of nitrosoaromatic ligands [75], as it is described in the Scheme 2.18.

The NO distances measured by X-ray diffraction analysis fall in the range 1.235–1.244 Å, what is the expected value for the N-adduct of the nitrosoaryl ligands (see Chap. 3).

2.6 Heteroaromatic Compounds

Heteroaromatic nitroso compounds are relatively rare, and their systematic study is available only in the recent literature. Taylor et al. [40, 76] prepared parent heterocyclic nitroso compound, 2-nitrosopyridine, and their derivatives, 3-methyl-2-nitrosopyridine and 4-methyl-2-nitrosopyridine by reaction of corresponding

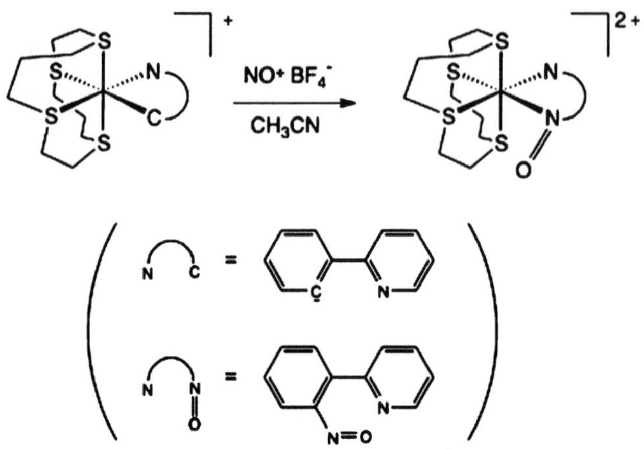

Scheme 2.18

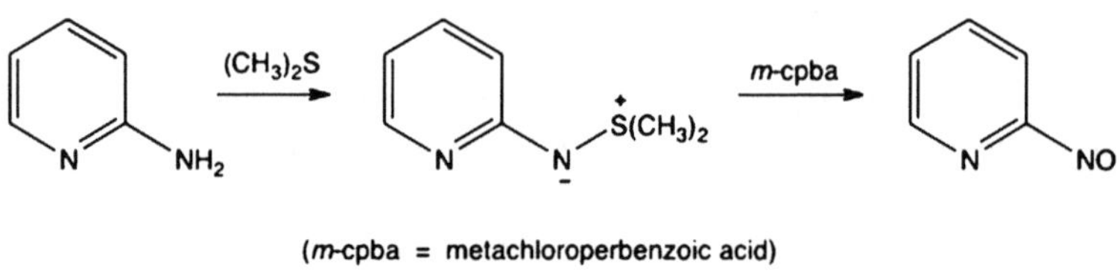

Scheme 2.19

aminopyridines with dimethylsulfide and *N*-chlorosuccinimide (Scheme 2.19). The resulted sulfonium salts were deprotonated to the dimethylsulfilimides and oxidized by *m*-chloroperbenzoic acid.

The obtained nitrosopyridine in its dimeric form can undergo hydrolysis (Scheme 2.20).

Nitrosopyrimidines can be easily synthesized by using the newly developed method by Marchal et al. [77]. A series of the nitrosoindolizine derivatives were prepared recently by Ghiviriga et al. [78].

Substituted nitroso-pyrazolines (Scheme 2.21) can be prepared upon the treatment of 1,3-diketones with sodium nitrite in acetic acid and hydrazines *via* formation of the oxime as an intermediate [79].

2.7 Nitrosoaromatic Compounds with More Nitroso Groups

Of three parent compounds, *o*-, *m*-, and *p*-dinitrosobenzene, the first is unstable because of its rearrangement to benzofuroxane, and the others readily polymerize with the mechanism that will be commented in subsequent chapters.

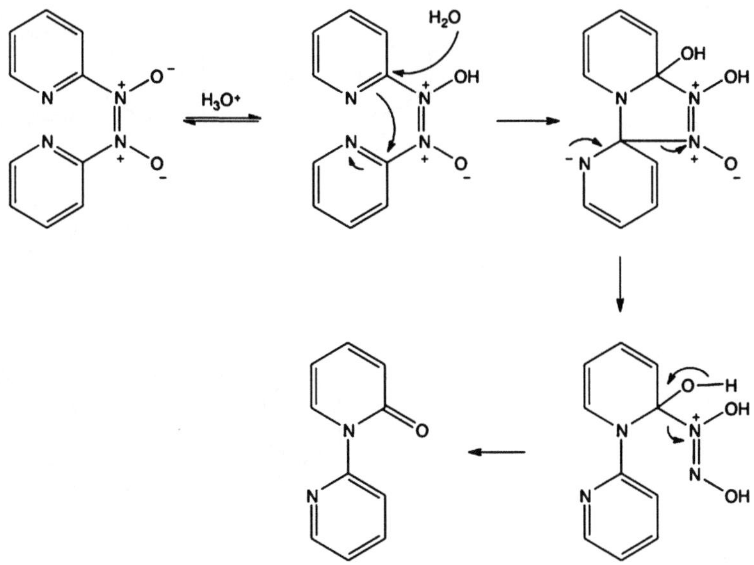

Scheme 2.20

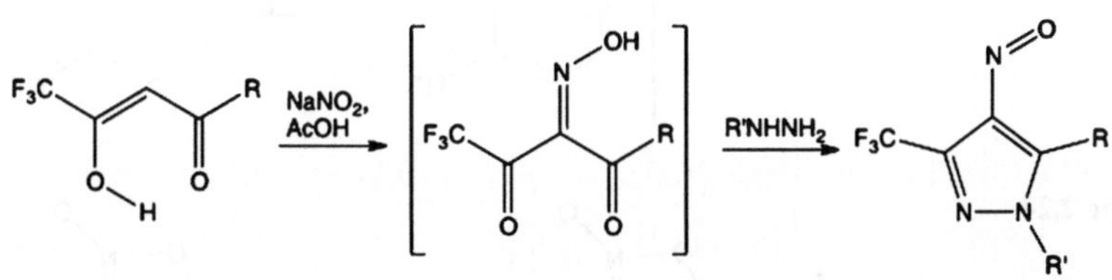

Scheme 2.21

Appearance of *o*-dinitrosobenzene by thermal rearrangement of benzofuroxan has in most cases been confirmed indirectly by trapping experiments [80–82]. Heating of benzofuroxan with *p*-anisyl azide yields diazoxy product, which is a result of the reaction of dinitrosobenzene with the *in situ* formed *p*-anisyl nitrene (Scheme 2.22) [82].

Hacker [83] succeeded to isolate 1,2-dinitrosobenzene in the argon matrix at 14 K and characterized it by IR and electron spectroscopy. Its N = O stretching band appears as an intensive absorption at 1,516 cm^{-1}. Of other characteristic signals the absorbances at 765, 790, 804, and 1,108 cm^{-1} were assigned to the C–N = O vibrations. In the UV spectrum the maximum has been found at 266 nm. Its blue shift relatively to the spectrum of parent nitrosobenzene (281 nm) could be explained by the contribution of the quinonoid resonance structure (Scheme 2.23).

The approaches to the synthesis of tetranitrosobenzene (benzodifuroxan) [84] and hexanitrosobenzene (benzotrifuroxan) [85] are known in the literature [86]. The hexanitrosobenzene – benzotrifuroxan rearrangement represents also an interesting model for studying aromaticity phenomena (Scheme 2.24).

Scheme 2.22

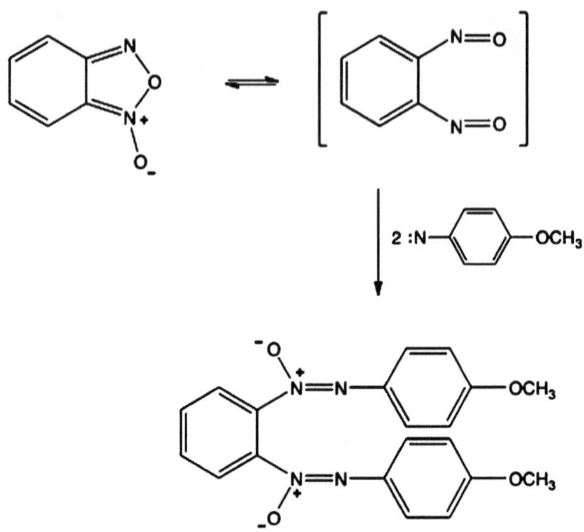

Scheme 2.23

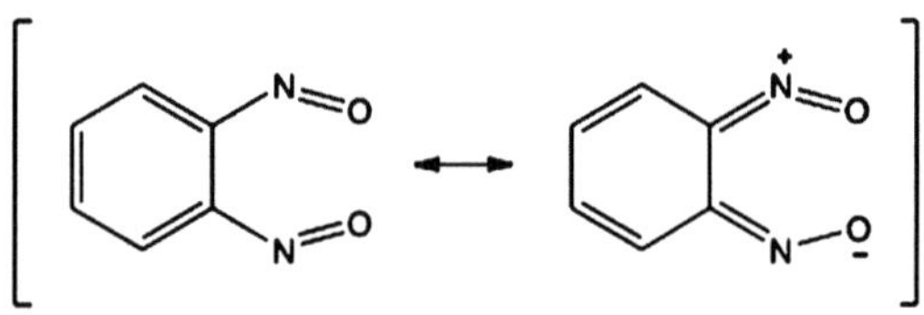

Scheme 2.24

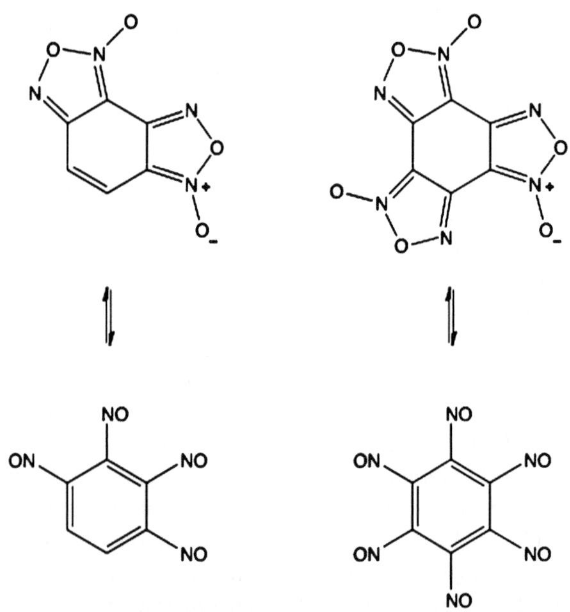

Scheme 2.25

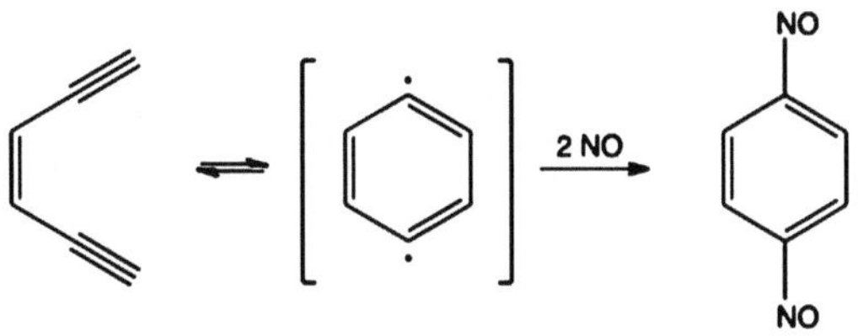

Scheme 2.26

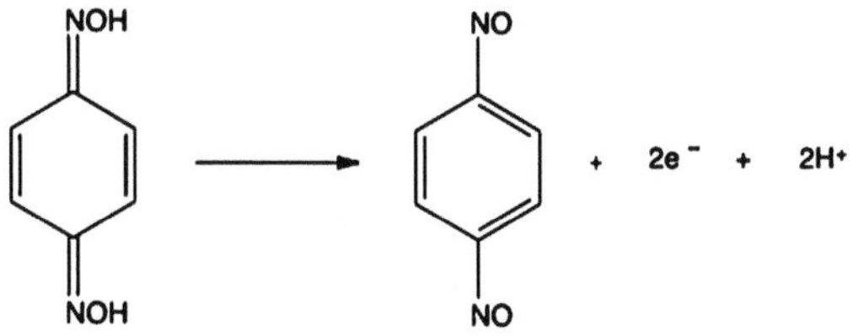

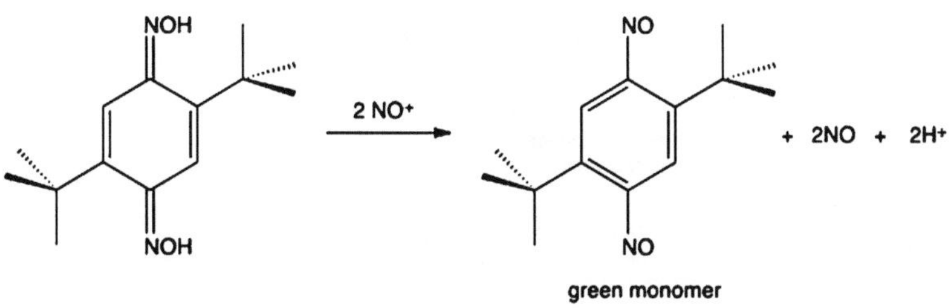

Scheme 2.27

For the preparation of *m*- and *p*-dinitrosobenzenes, the classical dihydroxylamine preparative route from the corresponding dinitro precursor can be used, analogously to the preparation of mononitrosobenzenes [87]. An alternative preparation of *p*-dinitrosobenzene includes the pyrolytic reaction of *cis*-hex-3-en-1,5-diyne with nitric oxide, *via* the formation of *p*-benzine (Scheme 2.25) [88].

However, Kochi et al. [89] have developed the most practical method, which is based on the autoxidation of benzoquinone dioxime (Scheme 2.26).

The preparation can be performed by the reaction with the nitrosonium ion, the origin of which could be either from nitrosonium tetrafluoroborate, or from disproportionation of NO_2 (Scheme 2.27). While the sterically hindered derivative yields green crystals with the monomeric structure, the substrate with the less voluminous *ortho* substituents leads to the formation of yellowish dimer, characterized with the IR absorption at 1,265 cm^{-1} (Scheme 2.28).

Instead of the nitrosonium cation, the chlorine has been used as an oxidant for the preparation of *p*-dinitrosobenzene polymers [90].

The analogous method has been successfully used for the preparation of 1,3,5-trinitrosobenzene (Scheme 2.29) [91]. The compound has been characterized on the basis of the intensive IR band at 1,272 cm^{-1}, assigned to the (O)N=N(O) stretching vibration of the dimeric (or oligomeric) product.

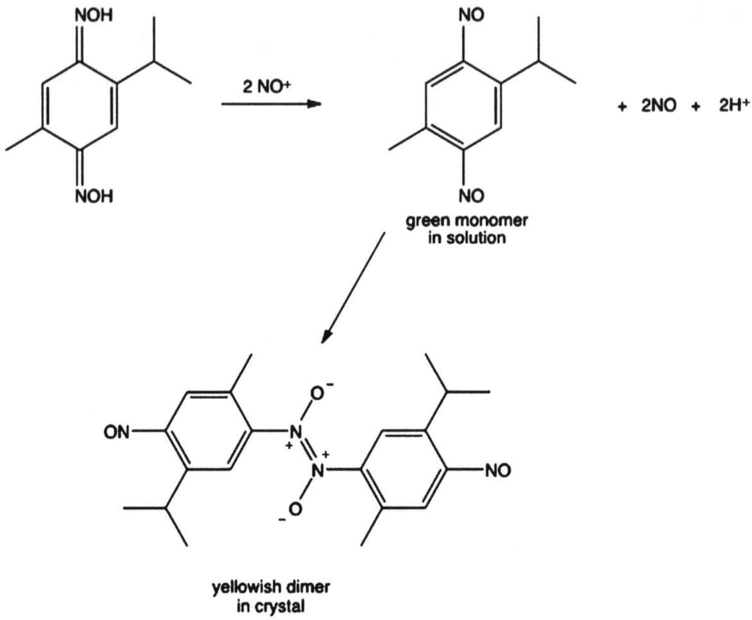

Scheme 2.28

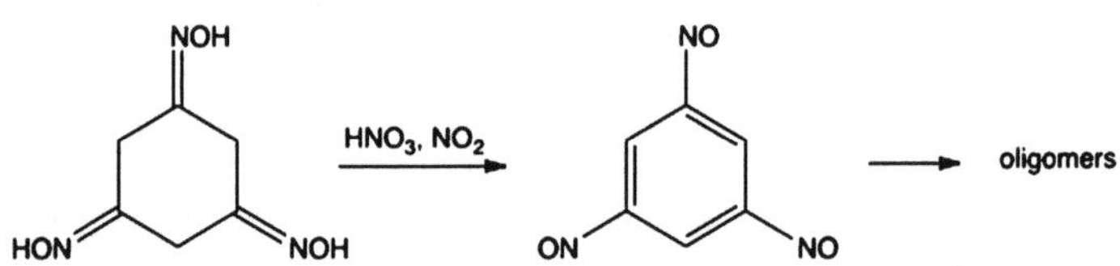

Scheme 2.29

2.8 Synthesis of Azoxides and Azodioxides

Azoxides belong to the most frequent side products on the synthetic red-ox route between limiting nitro- and aminoarene functionalities. Intermediate in this spectrum of structures, hydroxylamine, easily undergoes condensation with the nitrosobenzene to form azoxybenzene derivatives (Scheme 2.30).

Although appearance of azoxides constrains the efficiency of the preparative methods for the nitroso compounds, their potential applications based on the property to form liquid crystals could be a good reason to review this chemistry in more details. The earliest knowledge about the formation, as well as decomposition of azoxides could be dated to the beginning of the past century. Bamberger [92] and Knipscheer [93] discovered azoxybenzene derivatives as products of heating or pyrolyzing corresponding nitrosobenzenes. The mechanism of such thermal reactions was later explained by formation of the phenyl, and the NO· radical in the first step [94].

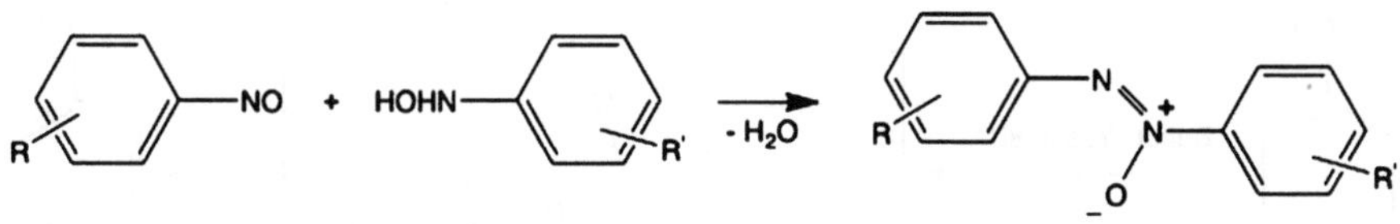

Scheme 2.30

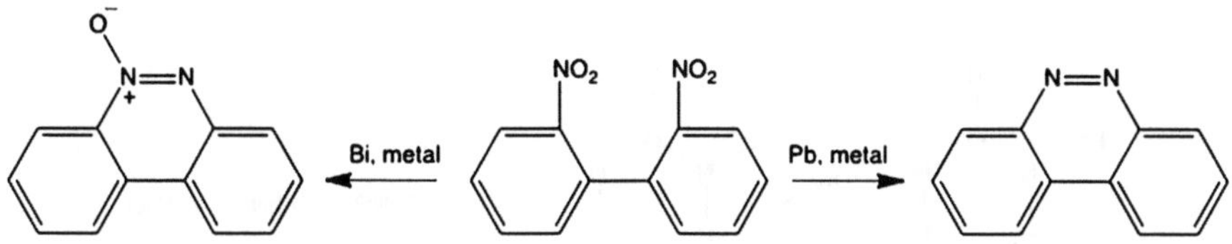

Scheme 2.31

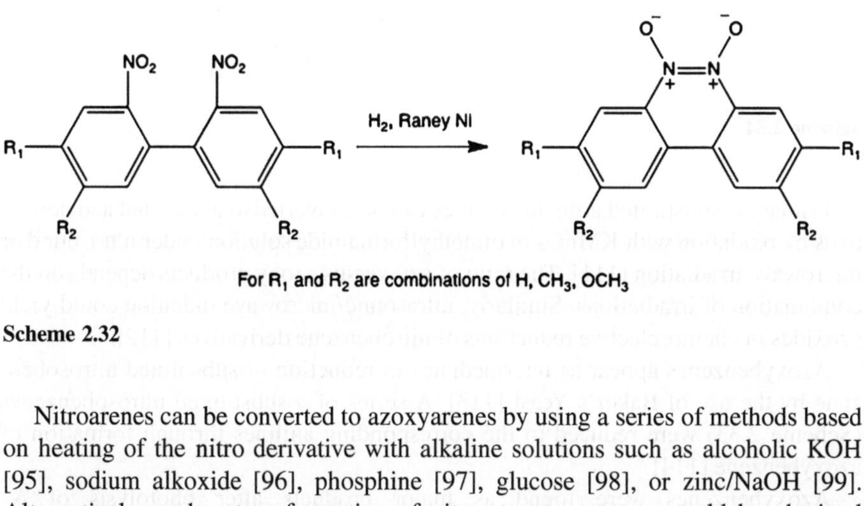

Scheme 2.32

Nitroarenes can be converted to azoxyarenes by using a series of methods based on heating of the nitro derivative with alkaline solutions such as alcoholic KOH [95], sodium alkoxide [96], phosphine [97], glucose [98], or zinc/NaOH [99]. Alternatively, such a transformation of nitro- to azoxyarenes could be obtained by standard methods of reductions with $LiAlH_4$, $NaBH_4$, or sodium arsenite [100–103]. It was found that especially efficient are reactions with metals, either as reductive reagents or as catalysts [104–106]. Solvent-free solid-state reactions of *p*-substituted nitrobenzenes with bismuth metal, yields substituted azoxybenzenes [107]. If 2,2′-dinitrobiphenyl is used as a starting compound, the product is cyclic azoxybiphenyl. However, replacing bismuth with lead affords formation of azo-derivative (Scheme 2.31).

Similar cyclizations were observed during reductions of differently substituted 2,2′-dinitrobiphenyles with hydrogen on Raney nickel [108] and with sodium bis(trimethylsilyl)amide [109]. Careful use of W-6 or W-7 Raney nickel as catalysts for hydrogenation, (Scheme 2.32) opened the opportunity to prepare also azodioxides [110].

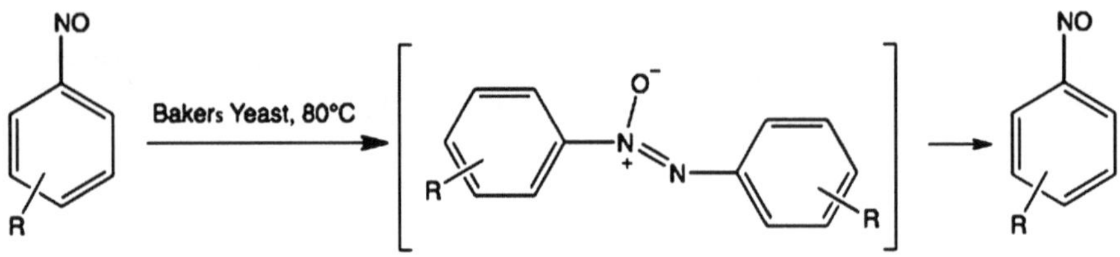

Scheme 2.33

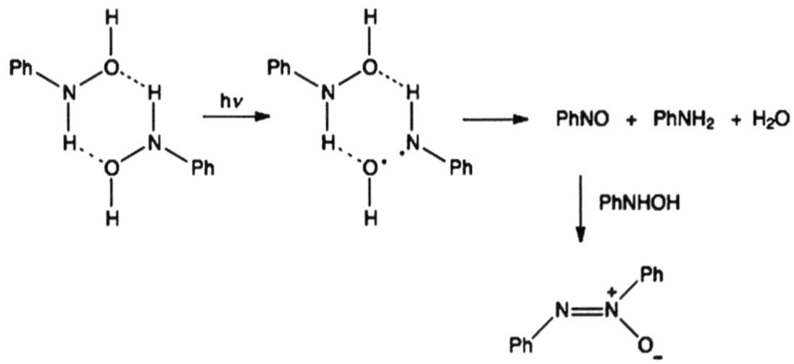

Scheme 2.34

Primary *p*-substituted aromatic amines can be converted to azoxy and azo derivatives by oxidation with $KMnO_4$ in dimethylformamide solution under ultrasound or microwave irradiation [111]. The ratio of azo versus azoxy products depends on the combination of irradiations. Similarly, ultrasound/microwave radiation could yield azoxides in chemoselective reductions of nitrobenzene derivatives [112].

Azoxybenzenes appear as intermediates in reduction of substituted nitrosobenzene by the use of Baker's Yeast [113]. A series of *p*-substituted nitrosobenzenes (Scheme 2.33) were reduced to the corresponding anilines through formation of azoxybenzene [114].

Azoxybenzenes were found as major products after photolysis of *N*-phenylhydroxylamine in acetonitrile [115]. Since the infrared spectra of the solution of *N*-phenylhydroxylamine showed bands assigned to the intermolecular hydrogen bonds between OH and NH groups [116], the reaction could begin from the hydrogen bonded dimer. The proposed mechanism (Scheme 2.34) that includes N–O cleavage is supported by findings that photolysis in crystalline state generates nitrosobenzene and aniline in almost equal amounts.

References

1. Baeyer A (1874) Chem Ber 7:1638–1640
2. Zharkova OM, Morozova YP, Trubacheva EN, Artyukhov VY (2009) Russ Phys J 52:458–463
3. Bartulin J, Belmar J, Gillardo H, Leon G (1994) J Heterocycl Chem 31:561

4. Belmar J, Jimenez C, Ortiz L, Garland MT, Baggio R (2006) Acta Cryst C 62:76
5. Enchev V, Angelova S (2009) J Mol Struct Theochem 897:55–60
6. Gowenlock BG, Richter-Addo GB (2004) Chem Rev 104:3315–3340
7. Maltha A, Favre TL, Kist HF, Zuur AP, Ponec V (1994) J Catal 149:364
8. Mars P, Krevelen DW (1954) Chem Eng Sci 3:41
9. Maltha A, Kist HF, Favre TLF, Karge HG, Asmussen F, Onishi H, Iwasawa Y, Ponec V (1994) Appl Catal A Gen 115:85
10. Zilberberg I, Ilchenko M, Isayev O, Gorb L, Leszczynski J (2004) J Phys Chem 108:4878–4886
11. Fletcher DA, Gowenlock BG, Orell KG (1997) J Chem Soc Perkin Trans 2:2201
12. Orell KG, Stephenskon D, Velarque JH (1990) J Chem Soc Perkin Trans 2:1297
13. Emery T, Neilands JB (1960) J Am Chem Soc 82:4903
14. Sklarz B, Al-Sayyab AF (1964) J Chem Soc:1318
15. Qureshi AK, Sklarz B (1966) J Chem Soc C:412
16. Maasen JA, de Boer J (1971) Recl Trav Chim 90:373
17. Bamberger E (1900) Bericht 33:1941; ibid 1953
18. Bartoli G, Leardini R, Medici A, Rosini G (1978) J Chem Soc Perkin Trans 1 1978:692
19. Bartoli G (1984) Acc Chem Res 17:109
20. Wrobel Z, Kwast A (2007) Synlett:1525–1528
21. Wirth S, Wallek AU, Zernickel A, Feil F, Sztiller-Sikorska M, Lesiak-Mieczkowska K, Bräuchle C, Lorenz IP, Czyz M (2010) J Inorg Biochem 104:774–789
22. Gard JC, Lessard J, Mugnier Y (1993) Electrochim Acta 38:677
23. Karakus C, Zuman P (1995) J Electrochem Soc 142:4018
24. Lamoureux C, Moinet C, Tallec A (1986) J Appl Electrochem 16:819
25. Lamoureux C, Moinet C (1988) Bull Soc Chim Fr:59
26. Gault C, Mionet C (1989) Tetrahedron 45:3429
27. Guilbaud-Criqui A, Moinet C (1992) Bull Soc Chim Fr 129:295
28. Vink JAJ, Cornelisse J, Havinga E (1971) Recl Trav Chim 90:1333–1336
29. Aurian-Blajeni B, Halmann M, Manassen J (1980) Solar Energy 25:165
30. Flores SO, Rios-Bernij O, Valenzueala MA, Córdova I, Gómez R, Gutiérrez R (2007) Top Catal 44:507–511
31. Brezová V, Tarábek P, Dvoranová D, Biskupic S (2003) J Photochem Photobiol A Chem 155:179
32. Blehert DS, Fox BG, Chambliss GH (1999) Bacteriol J 181:6254–6263
33. Hall M, Stuecker C, Ehammer H, Pointner E, Oberdorfer G, Gruber K, Hauer B, Stuermer R, Kroutil W, Macheroux P, Faber K (2008) Adv Synth Catal 350:411–418
34. Hall M, Stuecker C, Hauer B, Stuermer R, Friedrich T, Breuer M, Kroutil W, Faber K (2008) Eur J Org Chem:1511–1516
35. Toogood HS, Fryszkowska A, Hare V, Fisher K, Roujeinikova A, Leys D, Gardiner JM, Stephens GM, Scrutton NS (2008) Adv Synth Catal 350:2789–2803
36. Yanto Y, Hall M, Bommarius AS (2010) Org Biomol Chem 8:1826–1832
37. Mijs WJ, Hoeskstra SE, Ulmann RM, Havinga E (1958) Recl Trav Chim 77:746
38. Langley WD (1942) Org Synth 22:44
39. Holmes RR, Bayer RP (1960) J Am Chem Soc 82:3454
40. Taylor EC, Tseng CP, Rampal JB (1982) J Org Chem 47:552–555
41. Abramovich RA, Challand SR, Yamada Y (1975) 40:1541
42. Castellano JA, Green J, Kaufman JM (1966) J Org Chem 31:821
43. Sakaue S, Sakata Y, Nishiyama Y, Ishii Y (1992) Chem Lett:289
44. Tollari S, Cuscela M, Porta F (1993) J Chem Soc Chem Commun:1510
45. Porta F, Prati L (2000) J Mol Catal A 157:123
46. Zhu Z, Espenskon JH (1995) J Org Chem 60:1326
47. Möller ER, Jörgensen KA (1993) J Am Chem Soc 115:11814
48. Baldwin JE, Qureshi AK, Sklarz B (1969) J Chem Soc C:1073
49. Alizadeha MH, Tayebee R (2005) J Braz Chem Soc 16:108–111

50. Tundo P, Romanelli GP, Vázquez PG, Loris A, Aricó F (2008) Synlett:967–970
51. Bamberger E, Taschimer F (1899) Bericht 32:342
52. Priewisch B, Rück-Braun K (2005) J Org Chem 70:2350–2352
53. Crandall JK, Reix T (1992) J Org Chem 57:6759
54. Huskić I, Halasz I, Friščić T, Vančik H (2012) Green Chem
55. Gill A, Gandia LM, Vincente MA (2000) Catal Rev Sci Eng 42:145–212
56. Jagtap N, Ramaswamy V (2006) Appl Clay Sci 33:89–98
57. Yadav P, Sharma JK, Singh VK, Yadav KDS (2010) Biocatal Biotransform 28:222–226
58. Doerfe DR, Corbett MD (1991) Chem Res Toxicol 4:556–560
59. Kersten RD, Dorrestein PC (2010) Nat Chem Biol 6:636–637
60. Noguchi A, Kitamura T, Onaka H, Horinouchi S, Ohnishi Y (2010) Nat Chem Biol 6:641–643
61. Atherton JH, Moodie RB, Noble DR (1999) J Chem Soc Perkin Trans:699
62. Alkorta I, Garcia-Gómez C, de Paz JLG, jimeno ML, Arán VJ (1996) J Chem Soc Perkin Trans 2:293
63. Wirth S, Barth F, Lorenz I-P (2012) Dalton Trans 41:2176–2186
64. Pratt DS, Young CO (1918) J Am Chem Soc 40:1428–1431
65. Hou Y, Xie W, Janczuk AJ, Wang PG (2000) J Org Chem 65:4333–4337
66. Bosch E, Kochi JK (1994) J Org Chem 59:5573
67. Zyk NV, Nesterov EE, Khiobystov AN, Zefirov NS (1999) Russ Chem Bull 48:506–509
68. Atherton JH, Moodie RB, Noble DR (1999) J Chem Soc Perkin Trans 2:699
69. Bartlett EH, Eaborn C, Walton DRM (1970) J Chem Soc C:1717
70. Taylor EC, Danforth RH, McKillop A (1973) J Org Chem 38:2088
71. Molander GA, Cavalcanti LN (2012) J Org Chem 77:4402–4413
72. Birkofer L, Franz M (1971) Chem Ber 104:3062–3068
73. Bartlett EH, Eaborn C, Walton DRM (1970) J Chem Soc:1717–1718
74. Taylor EC, Danforth RH (1973) J Org Chem 38:2088–2089
75. Chan S-C, Cheung J-Y, Wong C-Y (2011) Inorg Chem 50:11636–11643
76. Taylor EC, Harrison KA, Rampal JB (1986) J Org Chem 51:101
77. Quesada A, Marchal A, Melguizo M, Nogueras M, Sánchez A, Low JN, Cannon D, Farrell DMM, Glidewell C (2002) Acta Cryst B B58:300–315
78. Ghiviriga I, El-Dien B, El-Gendy M, Martinez H, Fedoseyenko D, Metais EP, Fadli A, Katritzky AR (2010) Org Biomol Chem 8:3518–3527
79. Khudina OG, Burgart YV, Saloutin VI, Kravchenko MA (2010) Russ Chem Bull Int Ed 59:1967–1973
80. Kirby GW (1977) Chem Soc Rev 6:1–24
81. Gilchrist TL (1983) Chem Soc Rev 12:53–73
82. Bulacinski AB, Scriven EFV, Suschitzky H (1975) Tetrahedron Lett 16:3577–3578
83. Hacker NP (1993) Macromolecules 26:5937–5942
84. Boulton AJ, Gray ACG, Katritzky AR (1965) J Chem Soc 168:5958–5964
85. Bailey AS, Case JR (1958) Tetrahedron 3:113–131
86. Gowenlock BG, Richter-Addo GB (2005) Chem Soc Rev 34:797–809
87. Alway FJ, Gortner RA (1905) Bericht 38:1899–1901
88. Roth WR, Hopf H, Horn C (1994) Chem Ber 127:1765–1779
89. Rathore R, Kim JS, Kochi JK (1994) J Chem Soc Perkin Trans 1:2675–2684
90. Khishchenko YS, Makarov MA, Gareev GA, Cherkashina NA, Koptina GS (1969) J Appl Chem USSR (Eng Transl) 42:2245
91. Klyuchnikov OR, Khairutdinov FG (2004) Russ Chem Bull Int Ed 53:1133–1134
92. Bamberger E (1902) Bericht 35:1606
93. Knipscheer HM (1903) Recl Trav Chim Pays-Bas 22:1
94. Feinstein AI, Fields EK (1971) J Org Chem 36:3878–3881
95. Newbols BT (1961) J Chem Soc:4260
96. Suter CM, Dains FB (1928) J Am Chem Soc 50:2733–2739
97. Buckler SA, Doll L, Lind FK, Epstein M (1962) J Org Chem 27:794–798
98. Galbraith HW, Degering EF, Hitch EF (1951) J Am Chem Soc 73:1323–1324

99. Olah G, Pavlath A, Kuhn I (1955) Acta Chim Acad Sci Hung 7:71–84
100. Shine HJ, Mallory HE (1962) J Org Chem 27:2390–2391
101. Weil CE, Panson GS (1956) J Org Chem 21:803
102. Dewar MJ, Goldberg RS (1966) Tetrahedron Lett 7:2717–2720
103. Bigelow HE, Palmer A (1943) Org Synth Collect 2:57–59
104. Zechmeiser HE, Rom P (1929) Justus Liebigs Ann Chem 468:117–132
105. Hou Z, Fujiwara Y, Taniguchi H (1988) J Org Chem 53:3118–3120
106. McKillop A, Raphael RA, Taylor EC (1970) J Org Chem 35:1670–1672
107. Wada S, Mika U, Suzuki H (2002) J Org Chem 67:8254–8257
108. Kempter FE, Castle RN (1969) J Heterocycl Chem 6:523–531
109. Hwu JR, Das AR, Yang CW, Huang JJ, Hsu MH (2005) Org Lett 7:3211–3241
110. Kempter FE, Castle RN (1968) J Heterocycl Chem 5:583
111. Wu Z, Ondruschka B, Cravotto G, Garella D, Asgari J (2008) Synth Commun 38:2619–2624
112. Cravotto G, Boffa L, Bia M, Bonrath W, Curini M, Heropoulos GA (2006) Synlett 16:2605–2608
113. Tsuboi S, Sakamoto J, Kawano T, Utaka M, Takeda A (1991) J Org Chem 56:7177
114. Baik W, Rhee JU, Lee SH, Lee NH, Kim BH, Kim KS (1995) Tetrahedron Lett 36:2793–2794
115. White RC, Selvam T, Ihmels H, Adam W (1999) J Photochem Photobiol A: Chem 122:7–10
116. Mathise-Noel R, Munoz A, Mathis F (1961) Ann Fac Sci Univ Toulose Sci Math Sci Phys 25:113 (CA: 60: 10064g)

Chapter 3
Molecular Properties and Spectroscopy

This Chapter comprises the most important discussion about the aromatic C-nitroso compounds, their reactivity, typical reactions, photochemistry, electrochemistry, and spectroscopy. Since the most fascinating property of this class of the compounds is their monomer-dimer equilibrium, the basic molecular properties from the point of view of physical organic chemistry are represented. From the represented investigations of these phenomena, especially in the solid-state, it is demonstrated how these molecular species could serve as models for study of some important chemical concepts, solid-state reaction mechanisms, structure-selectivity relations, or photochromic systems.

3.1 Physical Organic Chemistry

Although in most scientific papers as well as in textbooks nitroso compounds does not serve as typical models for the study of basic concepts of physical organic chemistry, it was generally accepted that nitroso group as a substituent possesses unique properties, which can be helpful in enlightening some of basic ideas in the structure-properties relationships. Main investigation in this field was concentrated on the effects of the nitroso group as a substituent in aromatic reactivity studies, mechanisms of substitution reactions mediated by protonation of NO group, behavior of the aromatic nitroso compounds in superacid media, reactions in which nitrosobenzene derivatives serve as nucleophile, and mechanisms of the specific cycloadditions.

However, what makes physical organic chemistry of the nitroso compounds most interesting is the ability of the NO group to act ambivalently, as a nucleophile and as a target for nucleophile. The N-atom with its *n*-electrons behaves as nucleophile, and, because of the more electronegative oxygen in the neighborhood, it serves as a partially positive atom ready to accept electron pair from the nucleophile. Following these properties, let us develop this idea to discuss the most interesting consequence of such property, namely the strong tendency of the C-nitroso compounds to form

H. Vančik, *Aromatic C-nitroso Compounds*, DOI 10.1007/978-94-007-6337-1_3,

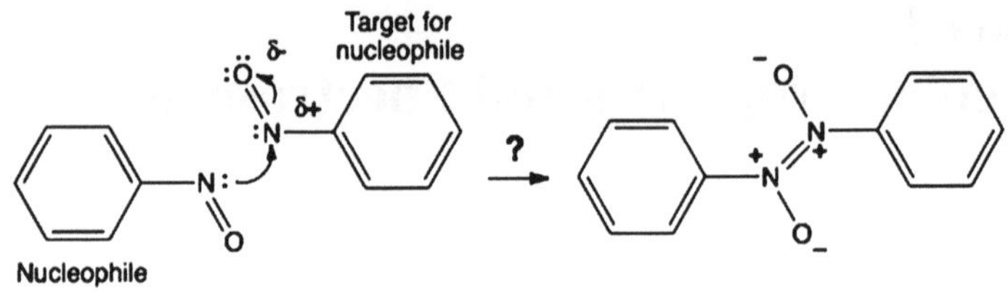

Scheme 3.1

dimers - azodioxides. Hypothetically, dimerization could perhaps be speculated as a nucleophilic "selfattack" (Scheme 3.1).

Probability of such a mechanism *versus* other two possibilities, single electron transfer, and a pericyclic-like concerted $2n + 2n$ electron addition, respectively, will be discussed later in this book. Let us start with some basic features of the aromatic nitroso molecules that play role not only in the nitroso chemistry, but also in physical organic chemistry in general.

3.1.1 Nature of the -NO Group

Exceptional nature of the nitroso group comes to light from a wide variety of experimental and theoretical approaches. It affords unique behavior as in the reactions that include a specific sort of dimerization, or the substituent effect in aromatic molecules, so in exhibiting special spectral patterns, especially in NMR spectroscopy. Consequently, nitroso compounds, and above all aromatic ones, reveal very wide variety of the reactions, which are applied in organic synthesis, crystal engineering, as well as in studying biological processes.

Perhaps, the best approach to the study of the nitroso group chemistry is investigation of the mechanism of the direct nitrosation of the benzene ring. In principle, the arene nitrosation and nitration bears mechanistic similarities because both include attack of either NO^+ or $NO_2{}^+$ electrophile. From the detailed study of the reaction mechanism, both the reactions follow similar pathway. Before the formation of the Whealand intermediate, nitrosation includes an additional mechanistic step, the electron transfer between the aromatic system and the NO^+ electrophile. The color change observed upon mixing of NO^+ with aromatic compounds has been attributed to the electron donor-acceptor π-complex (EDA) [1]. Persistence of such intermediates were later confirmed by quantum chemical calculations, spectroscopy, and from X-ray diffraction structures of analogs [2, 3] prepared from hexamethylbenzene and $NO^+AsF_6{}^-$ or $NO^+SbF_6{}^-$. The structure is shown in the Scheme 3.2.

Since the NO distances (1.13 or 1.36 Å) are closer to NO (1.15 Å) than to NO^+ (1.06 Å), the interaction with aromatic ring can be explained with a partial reduction of NO^+ caused by electron transfer from the benzene ring during formation of this EDA complex. The same conclusion also follows from the IR spectroscopy [3, 4],

Scheme 3.2

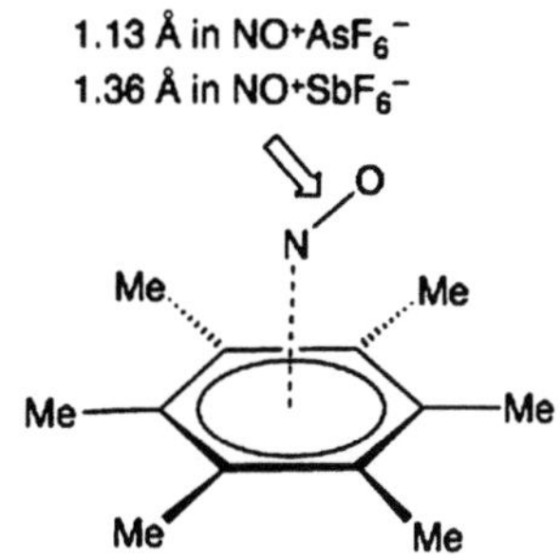

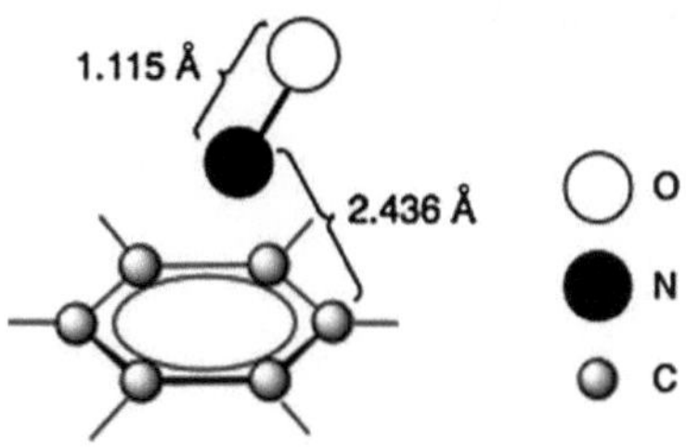

Fig. 3.1 Calculated geometry of the intermediate in the direct nitrosation of nitrosobenzene molecule with NO^+

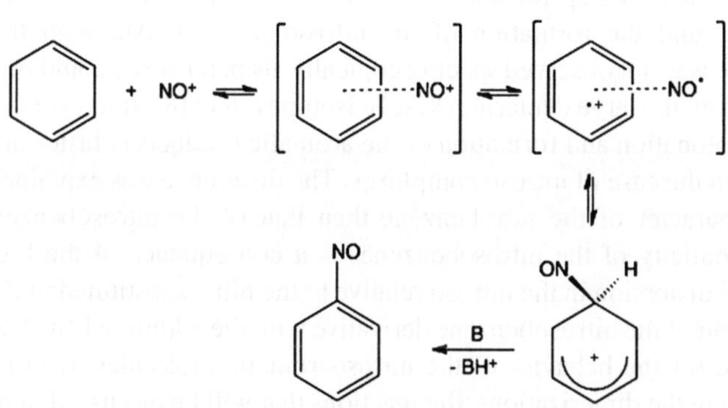

Scheme 3.3

while the NO stretching frequency in the EDA complex (1,849 cm^{-1}) resembles more to that of NO (1,876 cm^{-1}) than that of NO^+ (2,189 cm^{-1}). The increased cationic character of the aromatic ring also follows from the ^{13}C NMR spectra. From *ab initio* molecular orbital studies [5] follows the similar structure of the $C_6H_6 \cdots NO^+$ complex that is shown in Fig. 3.1.

Formation of the parent EDA complex *in situ*, i.e. in the nitrosation reaction has not been observed. Indications for the appearance of the π-complex come from the gas-phase spectroscopic investigations of the decompositions of $C_6H_6NO^+$ [6, 7].

Inclusion of the formation of EDA as an additional mechanistic step, proposed by Kochi et al. [8, 9], postulates a new general mechanism called *oxidative aromatic substitution*, represented in the Scheme 3.3.

Scheme 3.4

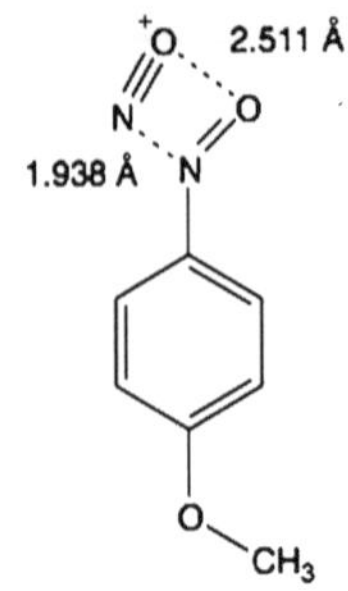

Scheme 3.5

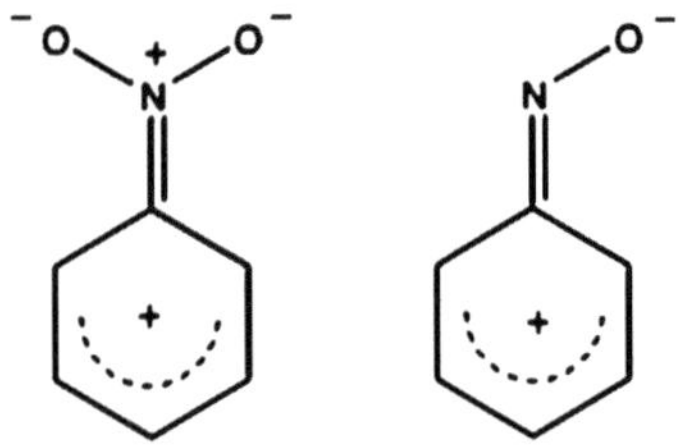

The last reaction step includes deprotonation of the Wheland intermediate by Lewis base, and the formation of the nitroso product. Although the Wheland intermediate was not observed spectroscopically, its persistence could be established indirectly from the large deuterium kinetic isotope effect [8]. Interesting observation is that deprotonation and formation of the aromatic products is faster in the case of nitro- than in the case of nitroso complexes. The difference was explained by higher aromatic character of the nitrobenzene than that of the nitrosobenzene product. Lower aromaticity of the nitrosobenzenes is a consequence of the higher degree of quinoidal distorsion in the nitroso relative to the nitro-substituted molecules [10].

Interaction of the nitrosobenzene derivative with the additional NO^+ electrophile is indicative for the behavior of the nitrosoaromatic molecules as in nucleophilic additions so in the dimerizations, the reactions that will be discussed in more details later in this book. Here, the nitroso N-atom behaves as a nucleophile: reaction of nitrosoanisole with nitrosonium hexafluorophosphate yields unstable adduct [10] in which the N-atoms of the nitroso derivate and nitrosonium ion appear in the vicinity. From the X-ray diffraction data obtained at −150 °C the N···N distance of 1.938 Å falls in between the single N-N bond (1.45 Å), and the sum of the van der Waals radii. The O···O distance is longer, 2.511 Å, which is for 0.529 Å less than the sum of the van der Waals radii (Scheme 3.4).

Classical method for selectivity-reactivity study in aromatic chemistry is based on the linear free energy correlation established by Hammett. While the effect of nitro group as a substituent on the aromatic molecule was systematically investigated [11], the influence of nitroso group was rather neglected in the literature. Both groups are known to be pronounced electron acceptors, which, consequently, deactivate benzene ring [12–16]. Such behavior was excellently explained by consulting their selected resonance formulas (Scheme 3.5).

Scheme 3.6

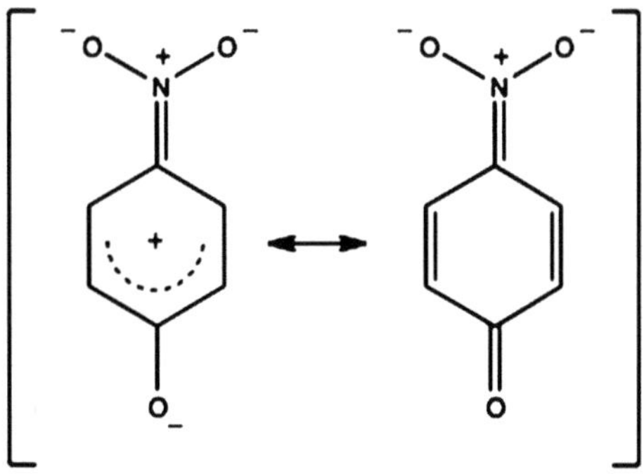

Looking on a total effect, nitro group behaves as a stronger electron withdrawing substituent because of the formal positive charge on the nitrogen atom. Deactivation of the benzene ring by -NO_2 must include not only resonance, but also a pronounced inductive effect [17]. Observation of its resonance interaction with the rest of the molecule is only possible if the additional, strong electron-donating group is bound in *para* position of the benzene ring. In *p*-nitrofenolate (Scheme 3.6), this resonance leads to high contribution of the quinoide structure [18].

In contrast, -NO group interact with the aromatic framework mostly by the resonance [19]. The difference between the two groups can be demonstrated by values of their σ_R and σ_I Hammett constants [20, 21]. While the measure for the resonance, σ_R, (0.33) of -NO is larger than in the case of -NO_2 (0.19), the inductive constant, σ_I, dominates in the nitro substituent (σ_I for -NO_2 is 0.56, while σ_I for -NO is 0.34). The more pronounced inductive contribution of -NO_2 in comparison with -NO is also reflected in dipole moments. Experimental dipole moment of nitrobenzene (3.97 D) is found to be larger than in the case of nitrosobenzene (3.20 D). Of the classical Hammett substituent constants [22], σ_p, -NO belongs to the upper limit with the highest value 0.91 (NO_2 is next on the list with $\sigma_p = 0.78$), and the σ_p^- parameter that specifically describes the resonance electron withdrawing influence is for -NO group 1.46. Because of this property, -NO was used as the best model for studying substituent effect on the electron redistribution in aromatic and antiaromatic systems.

Reactivity of the nitrosobenzene molecule, especially in nucleophilic addition reactions, is strongly influenced by *m*- or *p*- substituents on the benzene ring. Since the electron donors promote a "quinoide" resonance structure (Scheme 3.7), the nitrosobenzene molecule becomes less reactive as a target for nucleophilic attack. Details of such an influence of substituents on the degree of the quinoide contribution can be found in the systematic studies of the changes of the C-N and C-O bond distances in differently functionalized nitrosobenzene molecules. From the QSAR analysis [23] of a large series of substituted nitrosobenzenes, it follows that the most prominent electron acceptor, the NO_2 group, lengthens the C-N distance for 0.011 Å, and shortens the N-O distance for 0.004 Å, relatively to the parent nitrosobenzene. Oppositely, the dimethylamino group as the strong electron

Scheme 3.7

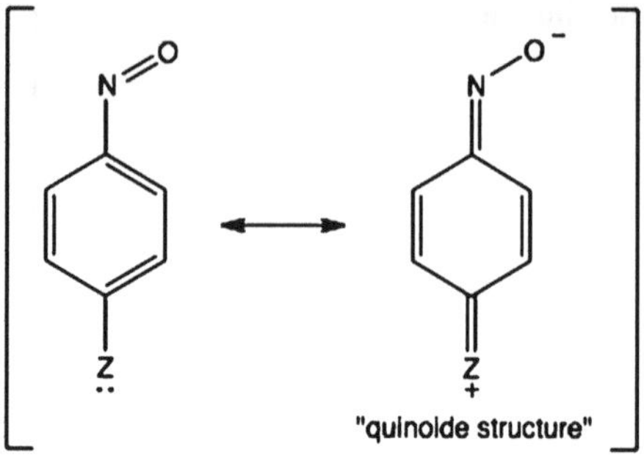

Table 3.1 Activation parameters for the nitroso group rotation relative to the benzene ring for the selected nitrosobenzene derivatives (Measured in CD_2Cl_2 or $CDCl_3$) (Data from Ref. [24])

Substituent	Solvent	$\Delta H^{\neq}$/kJ mol^{-1}	$\Delta S^{\neq}$/kJ mol^{-1}	$\Delta G^{\neq}$/kJ mol^{-1}
H	CD_2Cl_2	32.3	−7	34.3
4-F	CD_2Cl_2	35.1	−2	35.6
4-Cl	CD_2Cl_2	35.0	4	33.9
4-Br	CD_2Cl_2	34.3	1	34.0
4-I	CD_2Cl_2	33.7	−2	34.2
4-Me	CD_2Cl_2	39.2	12	35.7
4-OMe	$CDCl_3$	52.6	39	41.0
4-N(Me)$_2$	$CDCl_3$	60.3	25	52.8
4-N(Et)$_2$	$CDCl_3$	73.2	61	54.9
2-Me-4-NMe$_2$	$CDCl_3$	57.2	14	52.9
2,6-Me$_2$-4-NMe$_2$	$CDCl_3$	70.3	107	38.3

donor causes C-N shortening for 0.023 Å, and N-O lengthening for 0.009 Å. Later in this book we will discuss in more details how the same substituent effect plays fundamental role on the ability of nitroso monomers to form azodioxides.

Contribution of the quinoide structure has an effect on vibrational frequencies of N=O stretching mode, and strongly on the barrier of the restricted rotation of NO group around the C-N bond. Energies of barriers associated with the rotation of nitroso group have been measured by a temperature variable ^{1}H NMR bandshape analysis [24], and are represented in the Table 3.1. While the lowest $\Delta G^{\neq}$ values for the activation of the C-N rotation were measured for *p*-halogen, *p*-methyl, as well as for the parent nitrosobenzene molecule (34.0-35.7 kJ mol^{-1}, 298.15 K), *p*-amino substituents and *p*-alkoxy groups afforded the highest barrier (41.0-57.0 kJ mol^{-1}, 298.15 K).

The MP2/6-311G**//MP2/6-31G* calculated value 31.4 kJ mol^{-1} for the parent nitrosobenzene is comparable with these measurements [25]. In solid state the rotation barrier is only slightly higher than in the solution [26, 27]. The *n*-electron donors such as diethylamino and methoxy substituents enlarge the quinoide resonance effect, and as a consequence, corresponding nitrosobenzene derivatives do not dimerize to azodioxides. Reality of the explanation of such delocalization of electrons by resonance is additionally confirmed by finding of good correlation of these $\Delta G^{\neq}$ parameters with the Hammett σ^+ substituent constants [28].

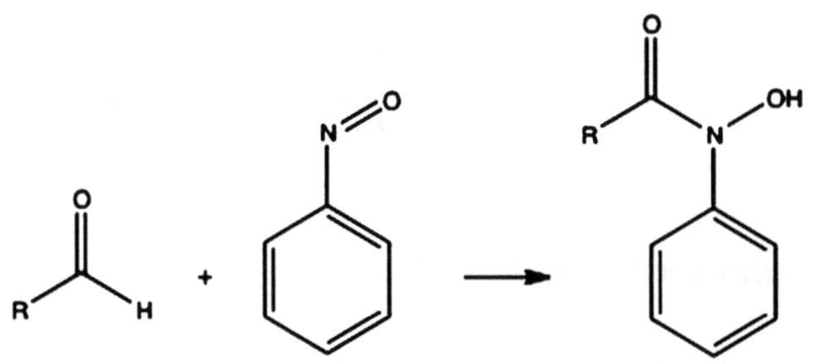

Scheme 3.8

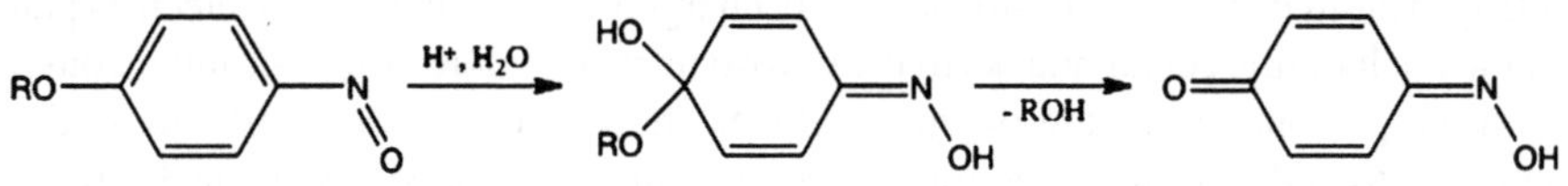

Scheme 3.9

3.1.2 Nucleophilic Properties

Since in -NO group two atoms possess nonbonding electrons, protonation could occur on the *N*- or on the *O*-position. Knowledge about the preference of protonation between two atoms is crucial for studying nitrosoaromatic derivatives as nucleophiles in nucleophile addition reactions. Detailed studies of the stability of protonated nitroso molecules have been made for pyrazole derivatives [29]. It was found that *O*-protonated derivative is for 16.4 kJ mol^{-1} more stable than the *N*-protonated one. In spite to that, there exist reaction mechanisms that include protonation of both, *N*- and *O*- centers of the nitroso group. Nucleophilic behavior in most of the reactions of *C*-nitroso compounds was explained by localization of the electron lone pair on the N-atom. It was experimentally supported by observation of the pronounced ^{15}N chemical shift anisotropy [30]. In addition reactions with carbonyl compounds, nitrosobenzene acts as a nucleophile molecule with its nitrogen atom as nucleophile reactive center. Uršić et al. have demonstrated that nitrosobenzene, as well as alifatic nitroso compounds, reacts with aldehydes forming *N*-phenylhydroxamic acids (Scheme 3.8) [31, 32].

It was found that the analogous reaction of nitrosobenzene occurs also with acyl chlorides and carboxylic acids [33, 34]. From the kinetic measurements including deuterium isotope effects follows that the proton transfer is the process included in the transition state [35–37].

Protonation of oxygen atom of the nitroso group seems to be a fundamental step in acid catalyzed hydrolysis of nitrosoaryl ethers. The reaction rates of different 4-nitrosoalkyl ethers in trifluoroacetic acid are strongly dependent on the acidity [38]. The proposed mechanism includes *O*-protonation as a first step. Such a protonated nitroso group strongly activates *para* (and *ortho*) positions in the benzene ring. By protonation, the nitroso group is converted to the oxime group in the reaction intermediate obtained after nucleophilic attack of the water molecule (Scheme 3.9). The structure resembles to the quinoide topology of the oxime analog.

Scheme 3.10

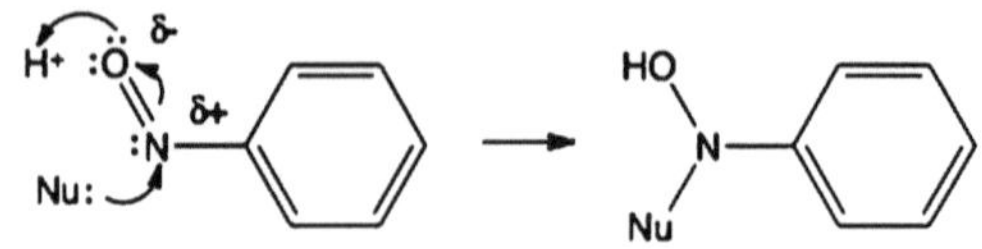

3.1.3 *Reactions with Nucleophiles*

Susceptibility of N = O group to addition of nucleophiles resembles analogous reactions of carbonyl group [39]. Because of the polarizability of N = O group, nitrogen atom becomes partially positively charged. In principle, a lone electron pair of the nucleophile center attacks to the nitroso nitrogen, more or less simultaneously with the protonation of the oxygen atom. Such class of reaction undergoes, also similarly to additions on carbonyl groups, general acid catalysis (Scheme 3.10).

Further development of the reaction depends strongly on the nature of the nucleophile, and on the catalytic conditions. However, there is also an alternative general mechanism, in which the single electron transfer appears as a crucial reaction step. In such case, reactants disproportionate to radicals or radical ions:

As we will discuss further in this book, both the general reaction mechanisms represent the conceptual basis for explanation of most important reactions of C-nitroso compounds, namely formations of azoxides and azodioxides (nitroso dimers).

3.1.3.1 Oxygen as Nucleophile

Addition of the water molecule yields the hydrated form of the starting nitroso derivative. However, in most cases its persistence was not directly confirmed by experiments, because of its high instability [40]. Hence, the equilibrium is shifted to reactants:

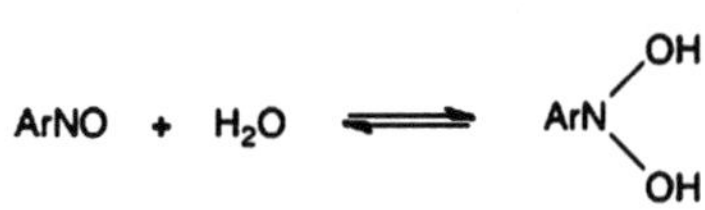

Observation of dihydroxyamine was only partially successful by using techniques of electrochemistry and spectroscopy during the investigations of reductions of nitroderivatives. Dehydration, which can be either base or acid catalyzed, is a very fast process, and determination of the lifetime of dihydroxyamine was only possible by using techniques such as pulse radiolysis [41]. By this approach, it has been measured that *p*-(dihydroxyamino) nitrosobenzene survives not more than 20 s at 20 °C [42].

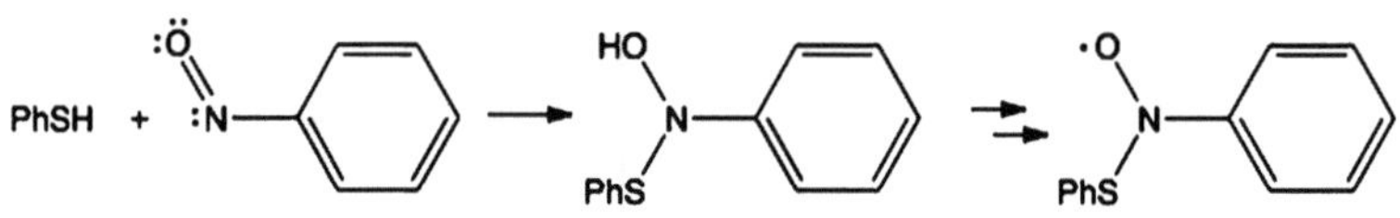

Scheme 3.11

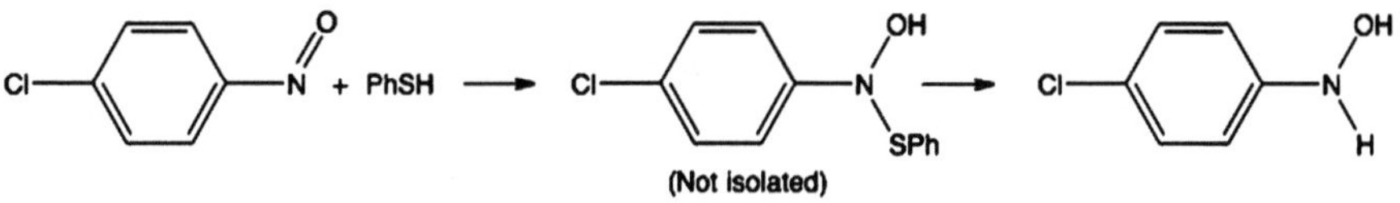

Scheme 3.12

Adducts obtained by reaction with alcoxide nucleophiles were not observed as yet. Indications for their persistence are only indirect, from studies of the reaction mechanism of azoxide formation.

3.1.3.2 Sulfur as Nucleophile

The earliest reports about the nucleophilic sulfur as an incoming group in the additions to the nitroso derivative deals with reaction of glutathione with 2-nitrosofluorene [43, 44], the process important for the study of the biological activity of nitrosoaromatic compounds. Investigation of simpler reaction, the addition of benzenethiol on the nitrosobenzene provided evidences that the attack of the thiol nucleophile is the first reaction step [45]. However, the detailed study of this reaction demonstrated that the complicated reaction mechanism, which includes single electron transfer as well as redox processes is taking a place. The appearance of nitroxide intermediate was confirmed by analyzing the *epr* spectral data. The mechanism is shown in the Scheme 3.11.

The extent and the rate of the initial step are sensitive to both, the substituent effect on the nitrosobenzene ring, as well as to the structure of the thiol. This conclusion follows from the systematic study of the product distributions after different thiol-nitrosoaromate reactions [46, 47]. From the *in situ* mechanistic studies, Paradisi et al. [48] have discovered that the mechanism strongly depends on the solvent used. Their proposed mechanism is shown in the Scheme 3.12. In alcoholic medium, the first step of the reaction involves formation of the unstable *N*-hydroxylbenzensulfenamide intermediate which, probably by breaking the N-S bond and abstraction of the H-atom from the environment, transforms to the hydroxylamine.

In the complex reaction scheme, after the interactions of these three components, the dominant isolated final products are azoxybenzene, sulfenamide and aniline derivative.

Sulfites readily react with nitrosobenzene forming the addition product. The rate-determining step is the nucleophile attack of the sulfite anion. Detailed

Scheme 3.13

Scheme 3.14

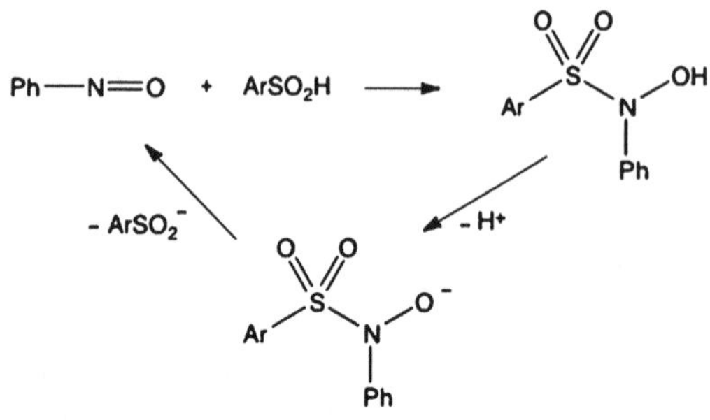

kinetic measurements demonstrated that the mechanism includes a single electron transfer with the formation of the radical pair [49]:

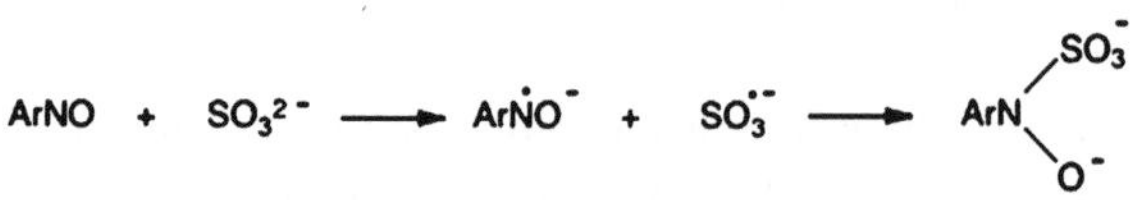

Anionic intermediate, as well as the anionic transition state is additionally confirmed by studying the Hammett correlations of differently substituted nitrosobenzenes. The positive ρ value of 2.65 has been found [40].

Especially interesting sulfur nucleophile is the sulfinic acid anion. Its electronic structure can be described by resonance forms, which are represented in the Scheme 3.13.

Nucleophilic attack of the sulfinic anion to the nitroso group is known after works of Bamberger et al. already from the nineteenth century [50]. Although the reactions of the sulfinic nucleophile with the aromatic nitroso compounds have been studied in the synthetic chemistry [51], reactivities as well as the stability of the obtained *N*-sulfonylhydroxylamine products were investigated only recently. If the nitrosobenzene has been used as the reactant, the resulted adduct remained unstable, since in the basic conditions the hydroxylamine molecule could be easily deprotonated, recovering the starting nitrosobenzene (Scheme 3.14).

However, Carroll and Conte [52] have discovered that 2-nitrosoterephthalic ester promptly reacts with the sulfinic anion yielding the stable cyclic derivative (Scheme 3.15).

Since sulfinic acid functionality appears as a product of the cysteine oxidation in physiological conditions, the reaction with this nitrosoaromatic molecule is proposed to be an analytical probe for the detection of these oxidative forms of cysteine. Convenience of such a method follows not only from the relatively short reaction time but also from the finding that the product is stable within the physiological pH range.

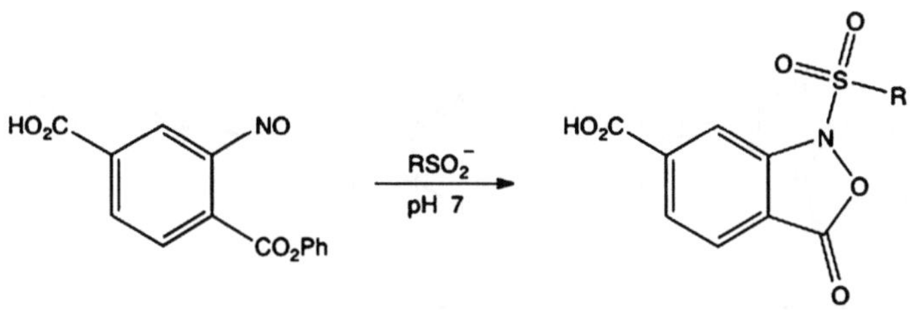

Scheme 3.15

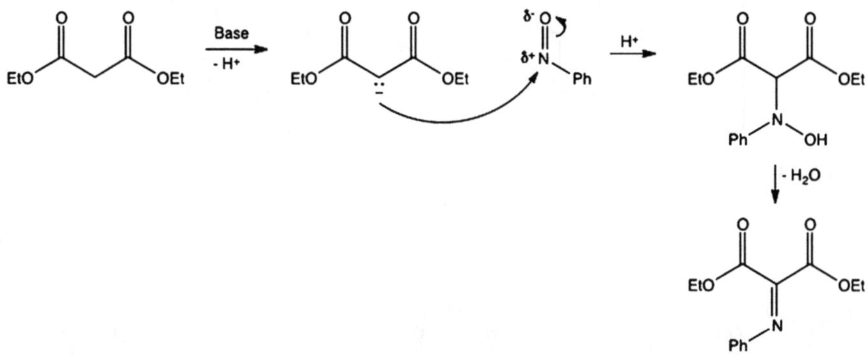

Scheme 3.16

3.1.3.3 Carbon as Nucleophile

Reaction of carbon nucleophile with the nitroso nitrogen is clearly demonstrated by the Ehrlich-Sachs type of the reaction [53] in which diethylmalonate anion serves as a nucleophile. The addition product is easily dehydrated to the azomethine derivative (Scheme 3.16). Such a product is predominant in the reactions with the substrates that yield more acidic carbon, such as phenylacetonitrile, malonitrile, or cyanoacetamide [54–57].

Enolate addition to the N-atom of the nitroso group appears as an essential step in the carbene catalysed annulation of enals and 4-nitrosotoluene, the reaction that is applied in the synthesis of seven-membered 4-azalactones [58]. Carbon nucleophile can be recognized as one of two resonance contributions to the structure of the intermediate (Scheme 3.17). Rearrangement to the final azalactone includes few additional reaction steps (Scheme 3.18).

Nitrosobenzene addition to the corresponding enolate is used as a crucial step in the decarboxylation of a series of functionalized cyclic and polycyclic aliphatic derivatives (Scheme 3.19). Such a nitrosobenzene-mediated synthetic procedures is of general importance because it affords products in excellent yields and in short reaction times [59].

The *N*-selective Aldol-type reaction of differently α-substituted aldehydes with nitrosobenzene catalyzed with organo-catalysts yields chiral product [60].

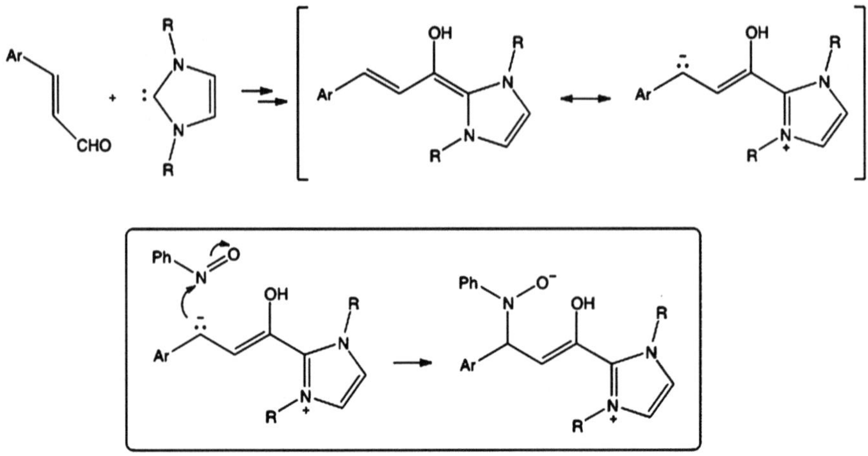

Scheme 3.17

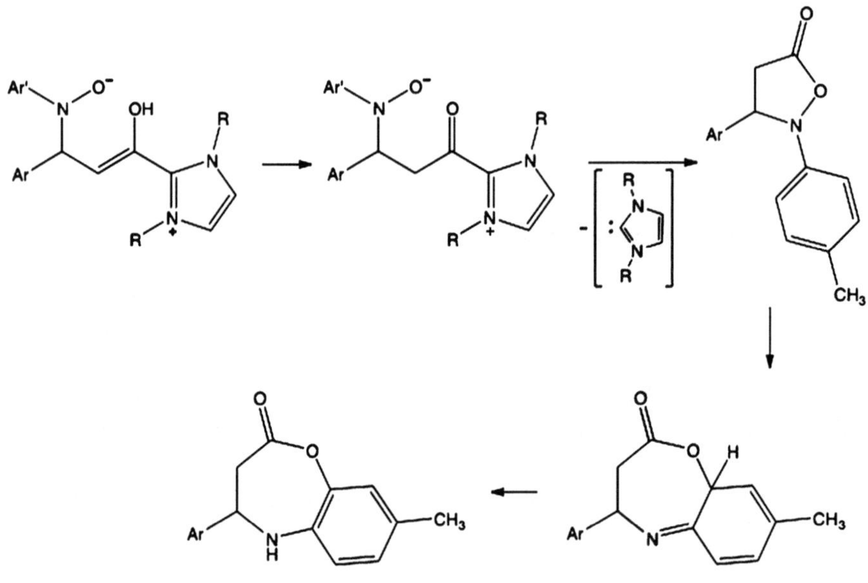

Scheme 3.18

The catalytic system includes a primary amine derived from the amino acid combined with 2,4-dinitrophenol. Again, the enantiomeric purity also depends on the solvent used (Scheme 3.20).

However, by using L-proline as a catalyst, the reaction is *O*-selective. Aminoxylation of a series of aldehydes proceeds with high enantiomeric purity (Scheme 3.21) [61, 62]. Disadvantage of this method is that a high excess (2-3 equivalent) of the carbonyl substrate is required for the good reaction yield.

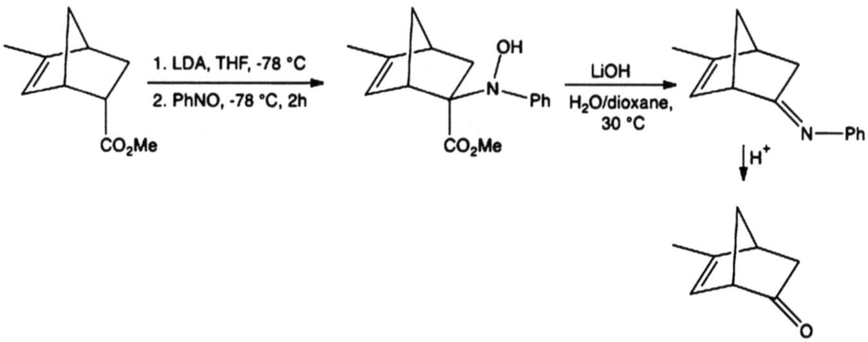

Scheme 3.19

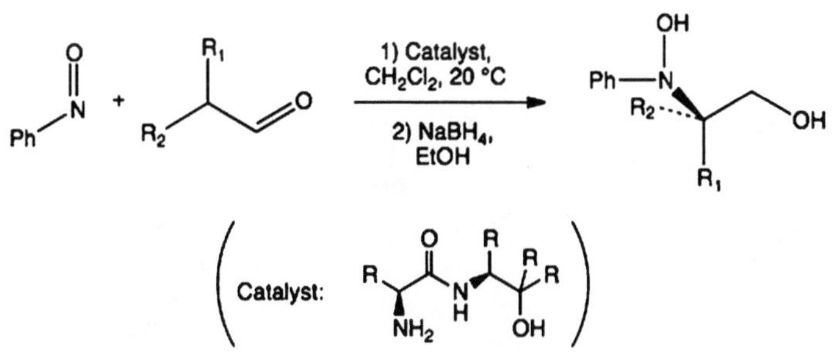

Scheme 3.20

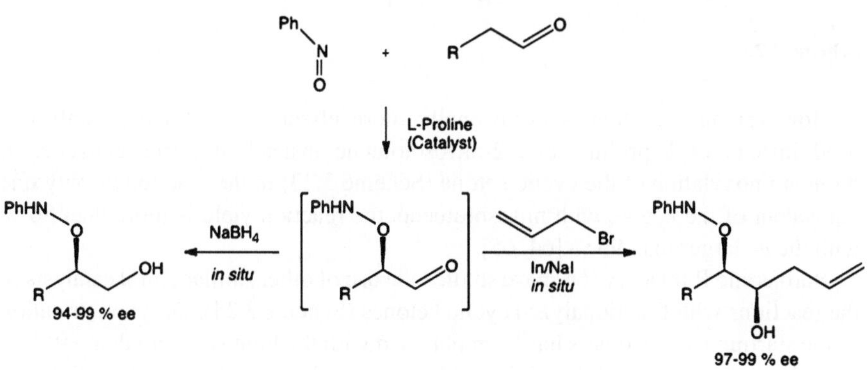

Scheme 3.21

After the first step in the reaction mechanism, which includes coupling of the substrate with the prolyne molecule, follows the addition of nitrosobenzene. MacMillan et al. [63] hypothesized that the enhanced Brønsted basicity of the nitrogen atom orients the nitrosobenzene molecule in such a way that the *O*-addition is preferred. The resulting reaction intermediate has a bicyclic spiro structure (Scheme 3.22).

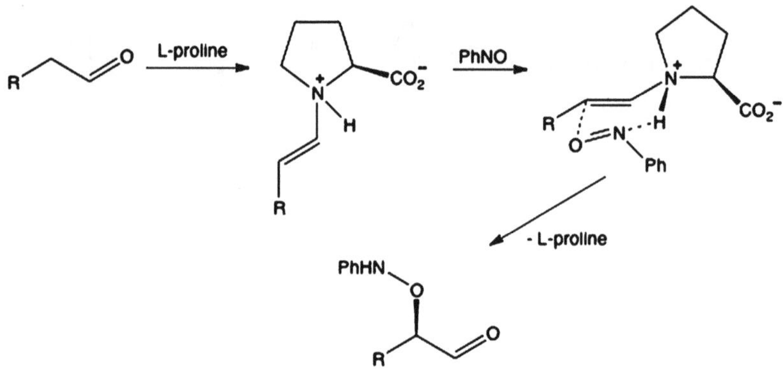

Scheme 3.22

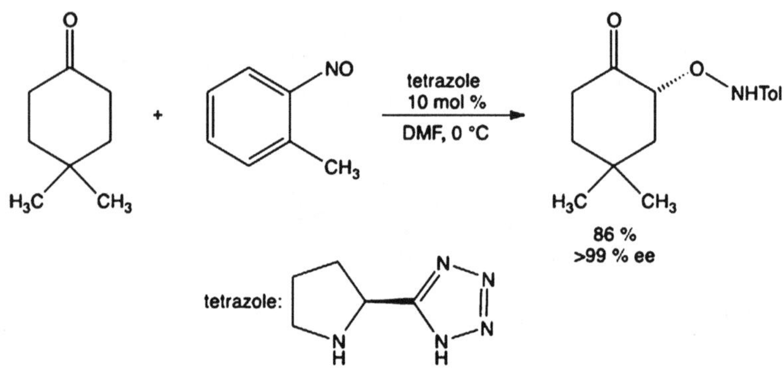

Scheme 3.23

However, the reaction is occasionally more effective if tetrazole catalyst is used instead of L-proline, and 2-nitrosotoluene instead of nitrosobenzene. In monoaminoxylation of the cyclic ketone (Scheme 3.23) in the reaction of only one equivalent of the ketone with nitrosotoluene, the reaction yield is more than 86 % with the *ee* larger than 99 % [64, 65].

Barbas and Ramachry [66] have studied the use of other similar chiral catalysts in the reactions with functionalyzed cyclic ketones (Scheme 3.24). Desymmetrization of the starting *meso*-isomers has been obtained with the high *ee*, more than 99 %.

If the carbon nucleophile originates from acidic hydrocarbons such as fluorene or indene, the nitrone appears as byproduct (Scheme 3.25). Its yield is strongly sensitive on the reaction conditions.

Cyanide ion reacts with the nitrosobenzene derivative analogously to its additions to aldehydes. In aprotic solvents, the primary adduct can be methylated to the stable product (Scheme 3.26) [67].

Nucleophilic addition of enol ethers to the N-atom of the nitroso group is recognized in recently discovered synthetic steps that include condensation of nitrosobenzene with silyl enol ethers (Scheme 3.27) [68, 69].

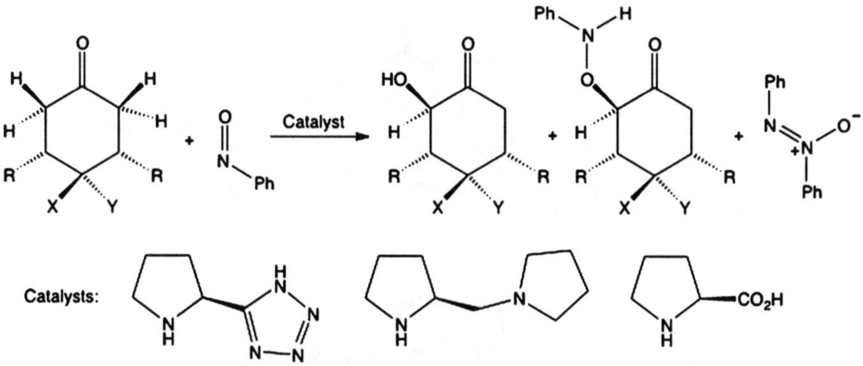

Scheme 3.24

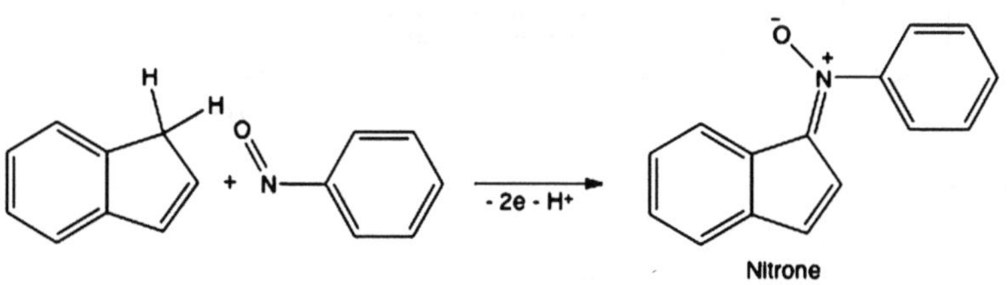

Scheme 3.25

Scheme 3.26

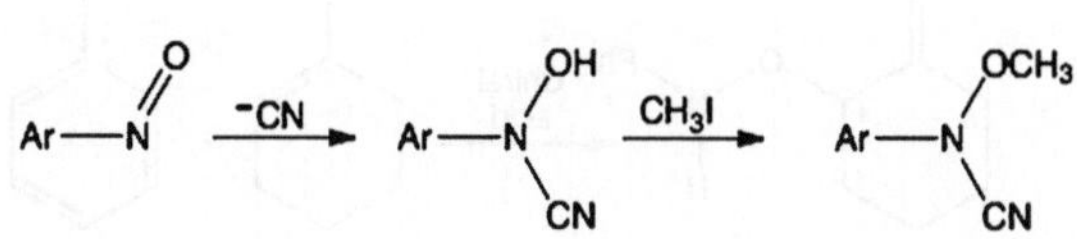

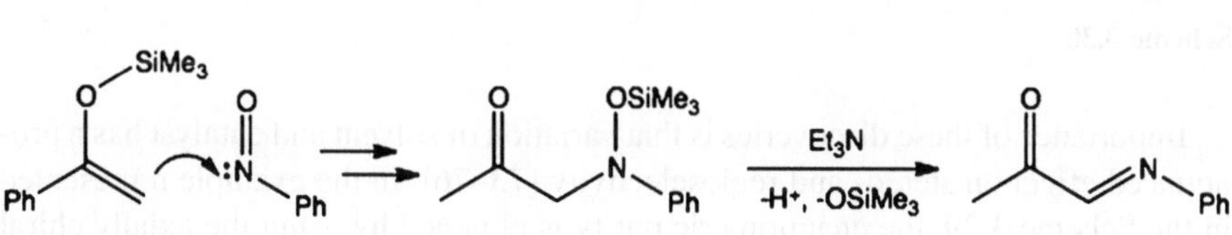

Scheme 3.27

Scheme 3.28

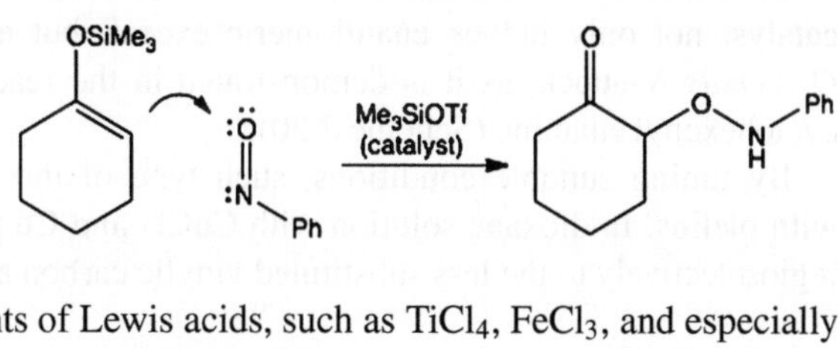

However, catalytic amounts of Lewis acids, such as $TiCl_4$, $FeCl_3$, and especially triethylsilyl triflate, promote nucleophilic addition to the O-atom [70–72]. In the reaction of silyl enol ether with nitrosobenzene, the α-aminoxy ketone was isolated as a sole product (Scheme 3.28).

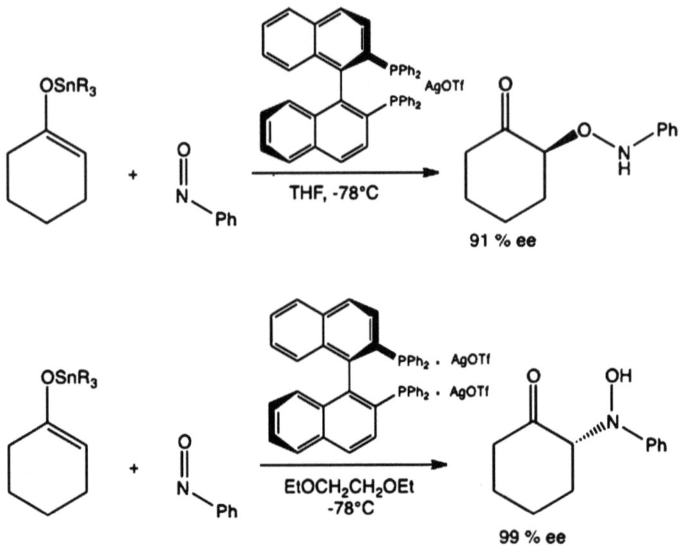

Scheme 3.29

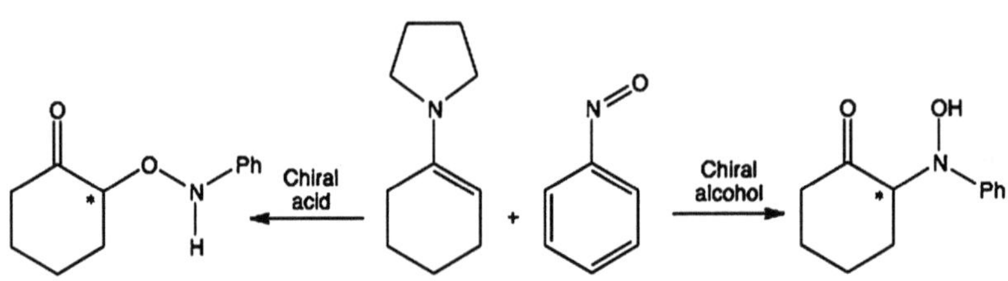

Scheme 3.30

Importance of these discoveries is that variation of solvent and catalyst has a pronounced effect on stereo- and regioselectivity [73–76]. In the example represented in the Scheme 3.29, the enantiomeric purity is obtained by using the axially chiral bisnaphtylphosphine derivative, and the *ee* highly depends on the solvent used.

Momiyama and Yamamoto [77] have discovered that selection of the chiral catalyst not only induce enantiomeric excess, but also direct the preference of *O- versus N*-attack, as it is demonstrated in the reaction of nitrosobenzene with cyclohexenyl enamine (Scheme 3.30).

By tuning suitable conditions, such type of the reaction could proceed also with olefins. In dioxane solution with $CuCl_2$ and Cu powder, nitrosobenzene adds regioselectively to the less substituted vinylic carbon atom [78]:

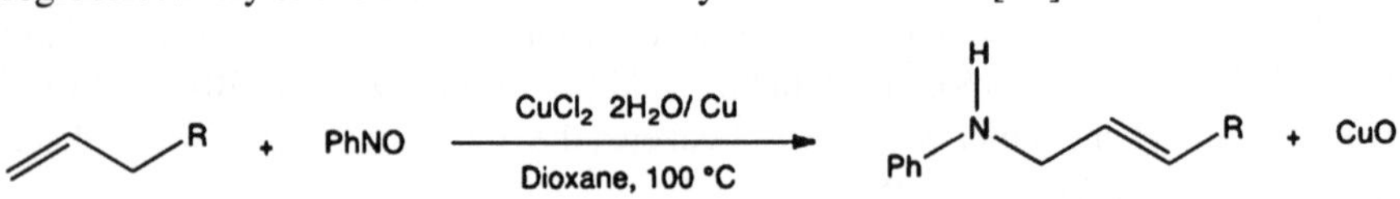

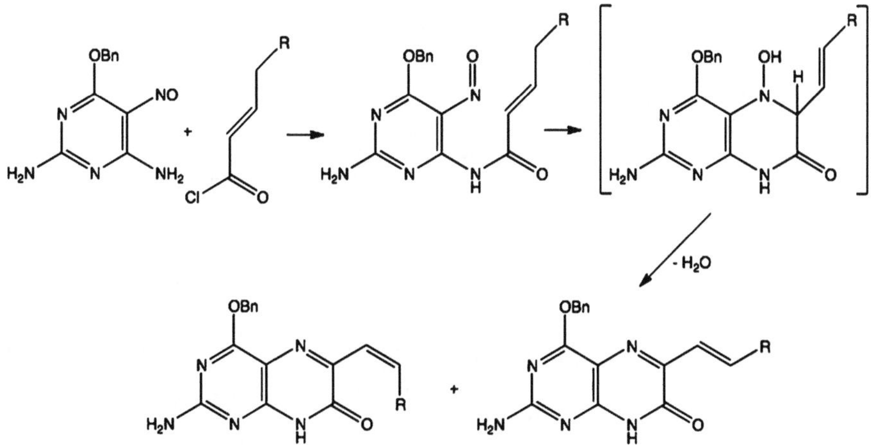

Scheme 3.31

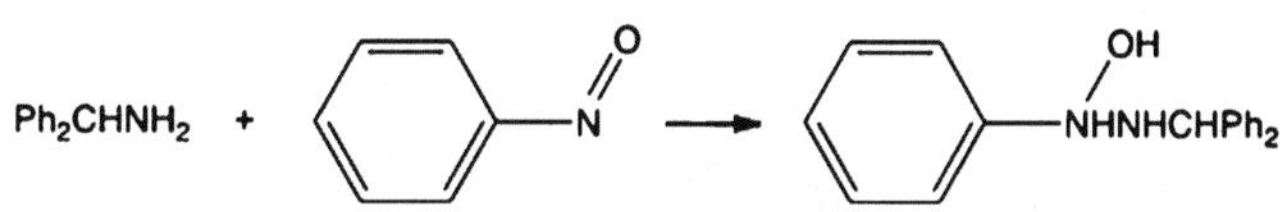

Scheme 3.32

Nitroso-ene addition can occur also by intramolecular mechanism. The reaction represents a valuable method for synthesis of natural products, as in the case of the construction of pteridine skeleton shown on the following scheme [79] (Scheme 3.31).

3.1.3.4 Nitrogen as Nucleophile

Knowledge about the reactions of nitrosoaromatic molecules with amines is important because the mixture of amines, hydroxylamines and nitroso compounds appears during the preparation of the nitroso compound by oxidation from the amine precursor. Namely, simultaneous appearance of these derivatives can yield azoxides and azo-compounds as byproducts. The amine molecule in its nonprotonated form reacts with the nitroso group by analogous mechanism as its addition to carbonyl group (Scheme 3.32).

The obtained adduct can by dehydratation be converted to azo-derivative or can decompose to imine and hydroxylamine (Scheme 3.33).

While primary amines with branched hydrocarbon skeleton prefer decomposition reaction, amines with straight hydrocarbon chains yields mostly azo derivatives [80–82]. Analogy with the addition to the carbonyl group could also be recognized in the reaction mechanism, i.e. both the reactions undergo general acid catalysis

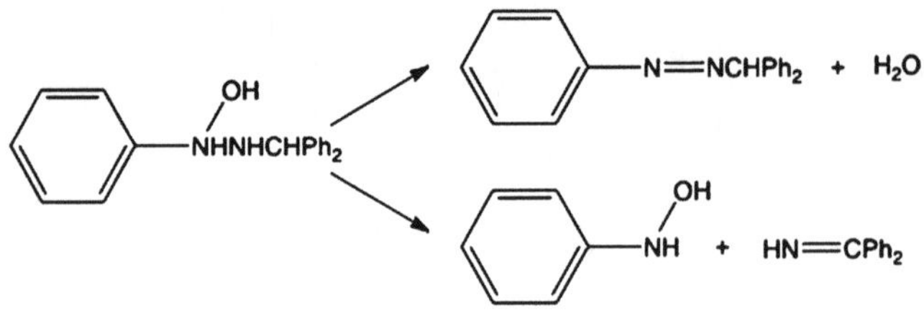

Scheme 3.33

Scheme 3.34

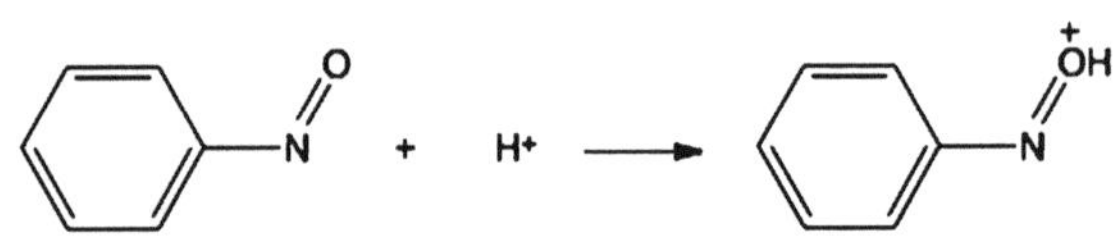

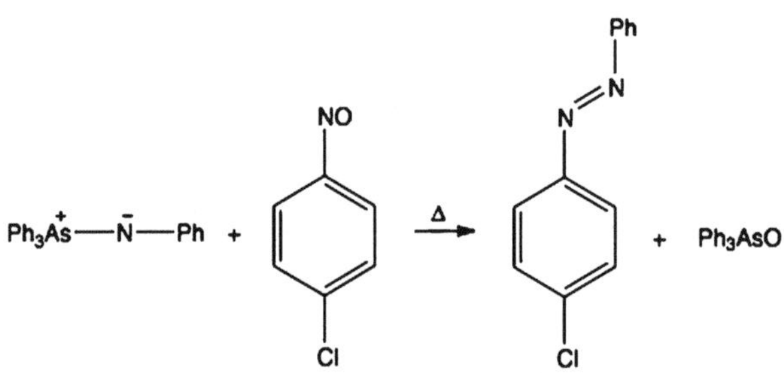

Scheme 3.35

(Scheme 3.34). In reaction of aniline with nitrosobenzene [83] derivatives the Brønsted α values were found to be 0.34. Consequently, the first step includes *O*-protonation of the nitroso group.

The structure of the transition state can be estimated from the Hammett equation data [84–89]. While variation of substituents on the aniline side follows correlation with the reaction constant $\rho = -2.14$, the change of substituents on the nitrosobenzene provides $\rho = +1.22$. Hence, reactivity of aniline derivative grows with its basicity, and in the transition state the positive charge develops on the nucleophilic nitrogen. Nitrosobenzene behaves oppositely, in the transition state the NO group becomes more negative and stabilized with the electron withdrawing substituents on the benzene ring. Since the absolute value of the reaction constant is larger for aniline, the efficiency of the reaction depends more on the electronic properties of nucleophile than on the electronic properties of the nitroso reactant. This evidence will be important in our explanation of the cross-dimerizations of differently substituted nitrosobenzenes, later in this book.

If instead of aniline, triphenylarsine phenylimine is used as a N-nucleophile, azo-compound appears as a product (Scheme 3.35) [90, 91].

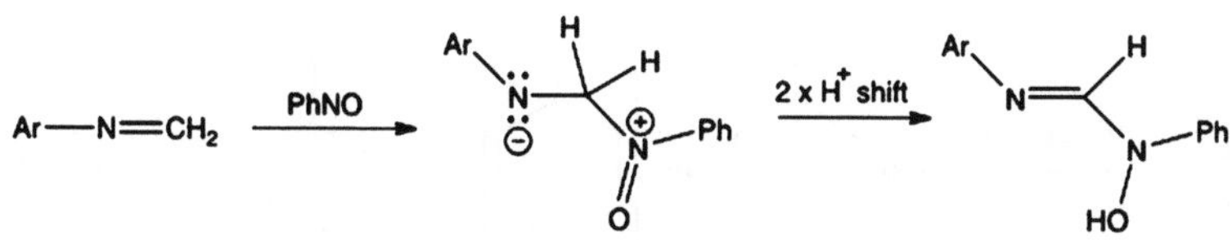

Scheme 3.36

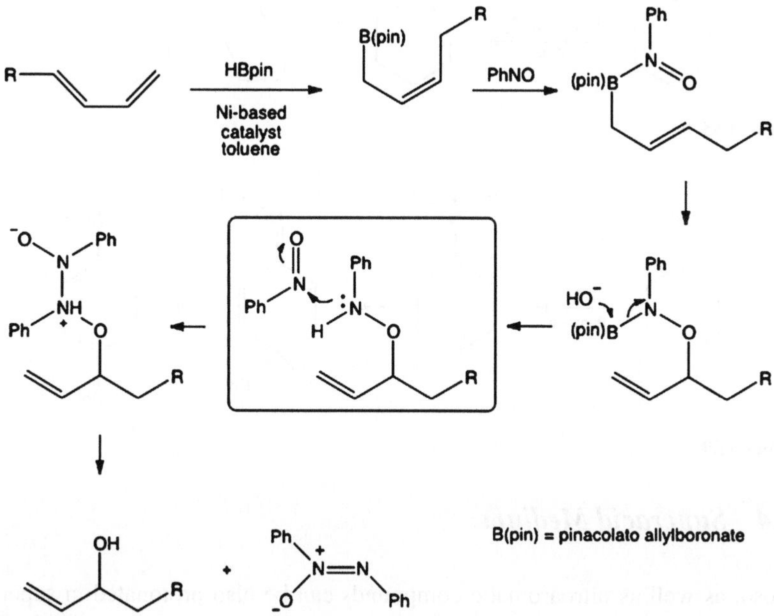

Scheme 3.37

In reactions with the molecules possessing highly electrophilic carbon, nitrosobenzene molecule attack by nucleophilic nitrogen producing the stable zwitterionic intermediate [92]. The structure of the Zwitterion has been confirmed by X-ray diffraction analysis. After dissolving in dichloromethane, two subsequent proton shifts led to the formation of the imino derivative (Scheme 3.36).

Study of the nucleophilic attack on the nitroso group by nitrogen is of fundamental importance because the dimerization of nitroso molecules probably follows such type of mechanism. An interesting example of the N-nucleophile reaction with the nitrogen atom of the nitroso group can be recognized in the intermediate reaction step during the allylation reaction of nitrosobenzene with pinacol allylboronates (Scheme 3.37). The complete reaction is convenient for the regioselective syntheses of allyl alcohols [93].

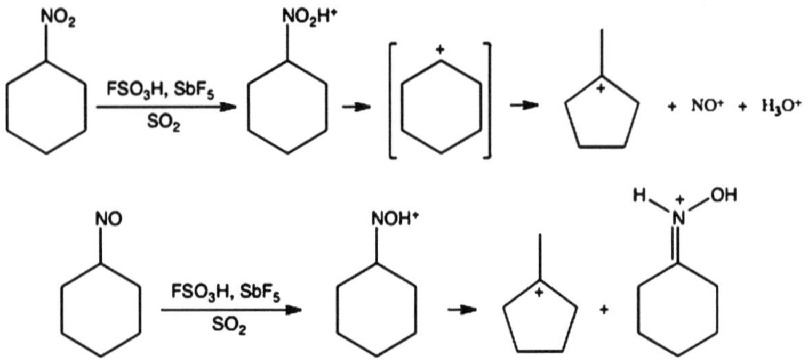

Scheme 3.38

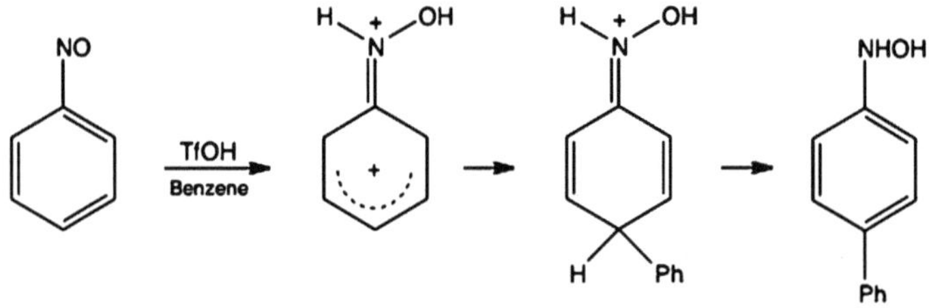

Scheme 3.39

3.1.4 Superacid Medium

Nitroso, as well as nitroaromatic compounds can be also protonated in superacid medium. Chemistry of superacids (magic acids), originally developed by G. Olah, enables broad investigations of stable carbocations by using NMR techniques [94]. Additionally, other experimental methods were applied in study of these important intermediates. Especially, investigations by IR spectroscopy introduced by H. Vančik [95] in combinations with high level quantum chemical calculations broadly used by P. v. R. Schleyer [96], opened opportunity to investigate details of the structure and effects operating in carbocations. However, protonation of nitroso and nitro compounds has special importance in chemistry because the resulting intermediates are frequently dications.

First observations of the intermediates obtained by protonation of nitro and nitroso derivatives were represented by Olah et al. [97, 98]. They have discovered that both classes of compounds after the *O*-protonation yield carbocations. While the nitro derivative releases nitronium ion as a byproduct, nitroso compound also forms an *N*-protonated oxime (Scheme 3.38).

Okamoto et al. [99, 100] have discovered that reaction of nitrosobenzene in trifluoromethylsulfuric acid (TfOH) in benzene solution produces *p*-biphenylhydroxylamine as the main product. The proposed reaction mechanism includes formation of the cation intermediate resulting by *N*,*O*-diprotonation of the nitroso group (Scheme 3.39).

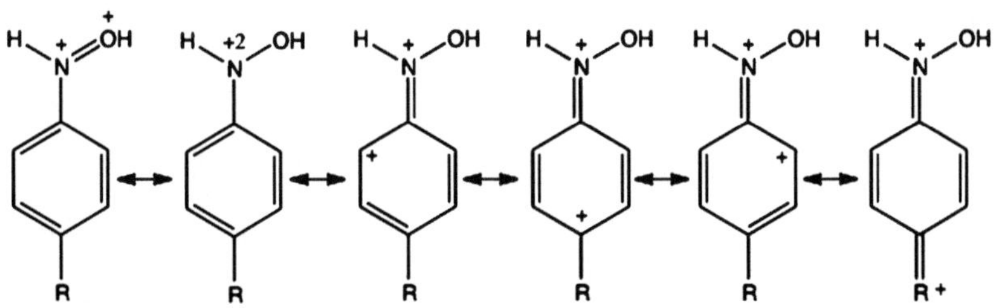

Scheme 3.40

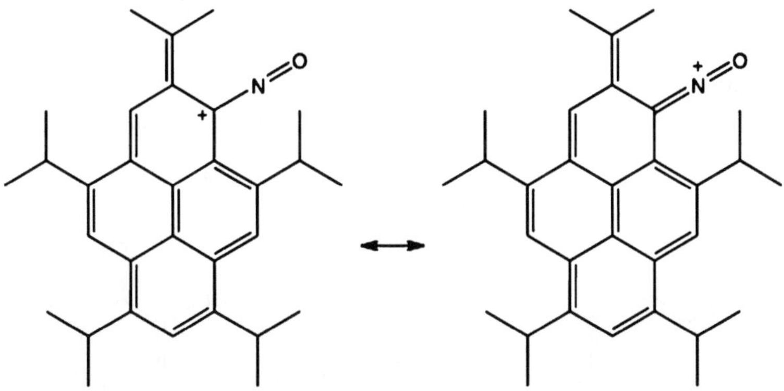

Scheme 3.41

Olah and Donovan [101] who demonstrated that the persistence of dications could additionally be stabilized by *n*-electron donating substituents in *p*-position (Scheme 3.40) have confirmed persistence of such species.

Diprotonation of the nitropyrene derivative by TfOH in SO_2 solution affords complicated set of the interconversions of dications. For discussion in this chapter, the most interesting seems to be formation of cationic species, which include nitroso group. The NO substituted carbocation is the example of the resonance stabilization by α-substituted nitroso group [102] (Scheme 3.41).

Unexpected persistence was observed for the nitroso radical cation (Scheme 3.42), which has been isolated as a one of the most stable product. The intermediate was characterized by spectroscopic methods including NMR and *epr* spectroscopies.

3.1.5 Nitroso Cycloadditions

Cycloadditions of nitrosobenzene are reactions of general use for synthesis of the cyclic skeleton in which oxygen and nitrogen are incorporated in the ring. Different types of cycloadditions have been reported, [4 + 2] with dienes, [3 + 2]

Scheme 3.42

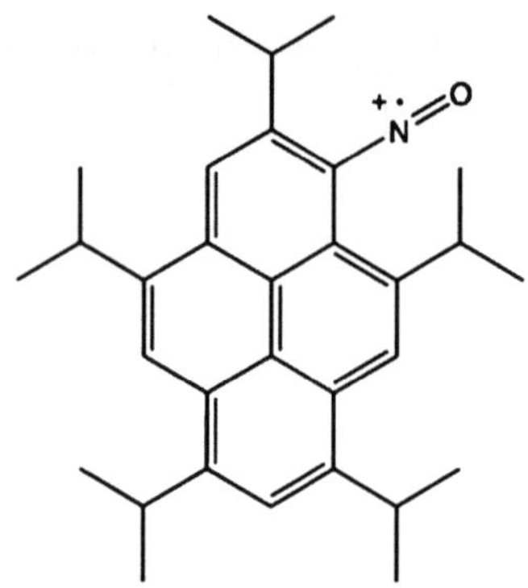

Scheme 3.43

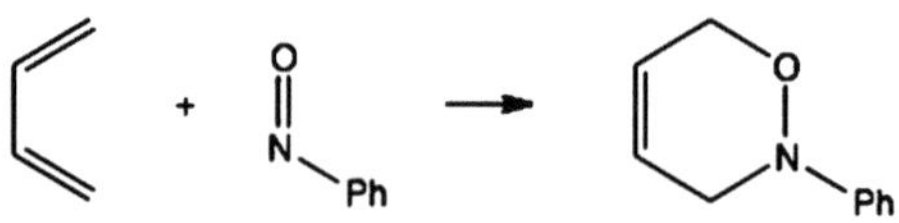

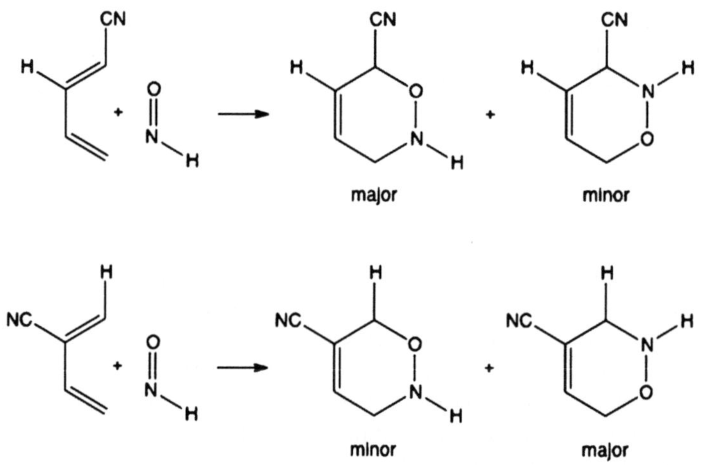

Scheme 3.44

with alkynes, and [3 + 3] or [4 + 2] with alkenylgold carbenoids. In the first observation of [4 + 2] addition (in the literature named as NDA, Nitroso Diels-Alder) by Wichterle [103] and Arbuzov [104] nitrosobenzene acts as a dienophile (Scheme 3.43). The 1,4-amino-oxo group is formed in a single step.

Diels-Alder cycloadditions of nitrosobenzene to the unsymmetrical dienes occur on the regioselective fashion. Addition of the nitrogen atom to the more substituted diene carbon depends on the nature of the substituent on the diene molecule. In the calculated model systems where HNO is used instead of the nitroso molecule, the positions and electron withdrawing ability of the diene substituents define the regioselectivity (Scheme 3.44) [105].

Scheme 3.45

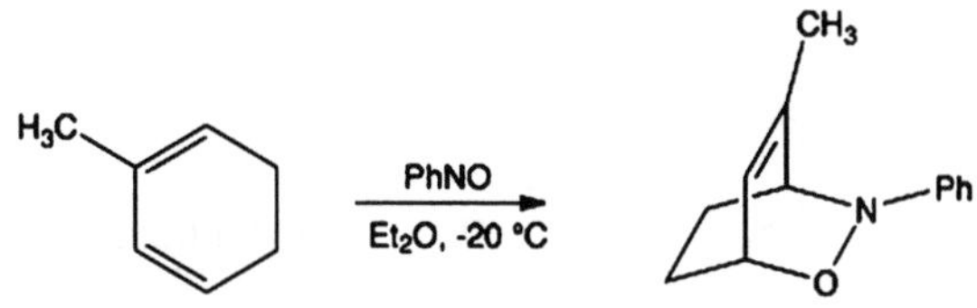

Scheme 3.46

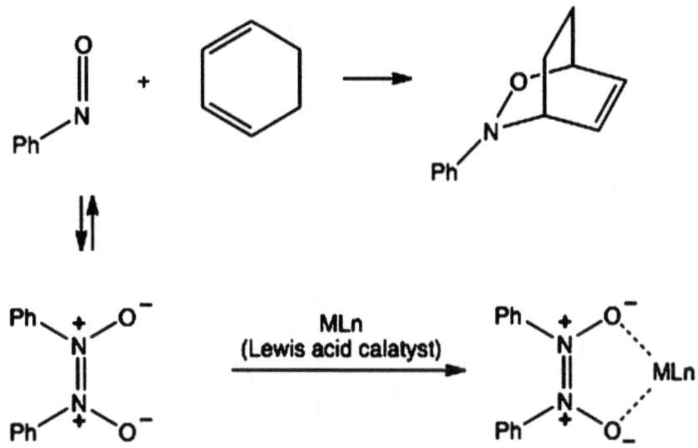

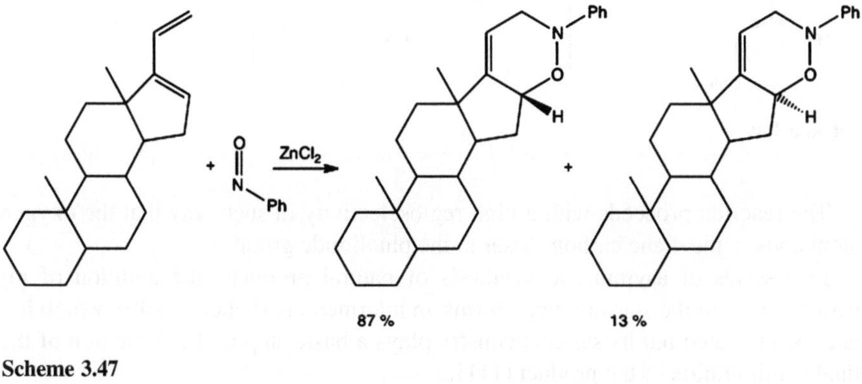

Scheme 3.47

Such type of a regioselective reaction is preferred in the case of the 2-methylbutadiene structure (Scheme 3.45) [106, 107].

The reactions are in most cases more efficient in the presence of Lewis acid metal catalysts. However, since nitrosobenzene in solution appears also as *Z*-dimer (azodioxide), it frequently forms complex with the catalyst (Scheme 3.46), and in such case the Lewis acid losses its catalytic ability [108].

As former scheme shows, it could not be expected that Lewis acid enhance the reaction. On the other hand, Lewis acid acts on the enantioselectivity. In the presence of $ZnCl_2$, the NDA addition to the steroid represented on the Scheme 3.47 yields 87 % excess of one of the isomers [109].

Reactions of nitrosobenzene with seven-membered 6-phtalimidobicyclo[5.1.0] octa-2,4-diene in methylene dichloride gives the corresponding cycloadduct (Scheme 3.48) [110].

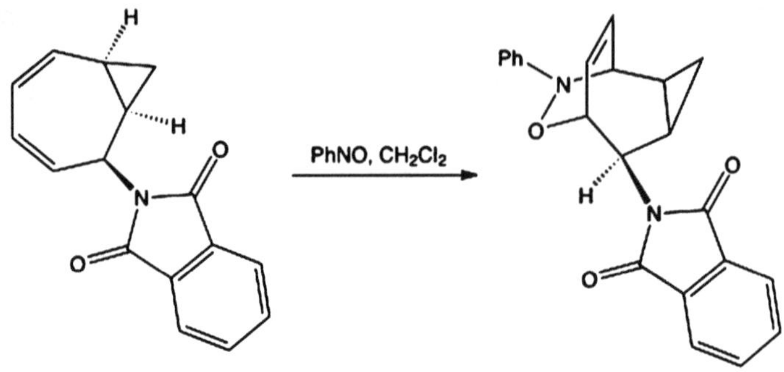

Scheme 3.48

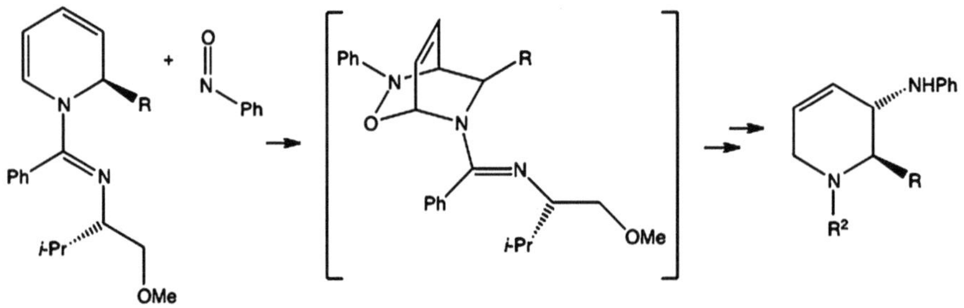

Scheme 3.49

The reaction proceeds with a high regioselectivity in such way that the oxygen atom adds to the diene carbon closer to the phtalimide group.

In a series of asymmetric synthesis of natural products, the addition of nitrosobenzene to the starting diene forms an intermediate (Scheme 3.49), which has not been isolated but its stereochemistry plays a basic step in the formation of the final configuration of the product [111].

Even without the catalyst, the NDA reaction affords a high regioselectivity as in the case of the 5-dienyl pyrimidone (Scheme 3.50) [112].

Depending on the reaction conditions, adduct of NDA can also undergo retro Diels-Alder reaction. The tetramethylpurpurogallin natural product reacts in hexane solution with nitrosobenzene forming the stable adduct (Scheme 3.51). However, in different solvents such as diethyl ether or methylene dichloride the adduct dissociates to the reactants by retro D-A reaction [113].

In combinatorial chemistry, the NDA reactions can be performed with aromatic nitroso derivatives immobilized on the polymer resin. Nitrosopyridine molecules fixed for the *Rink amide* resin serve as efficient dienophiles for a series of dienes, some of which are important natural products (Scheme 3.52) [114].

The specific stereochemical conditions on the solid surface direct the stereochemistry of the products. It was found that a favorable spatial arrangement plays a critical role for fast reactivity.

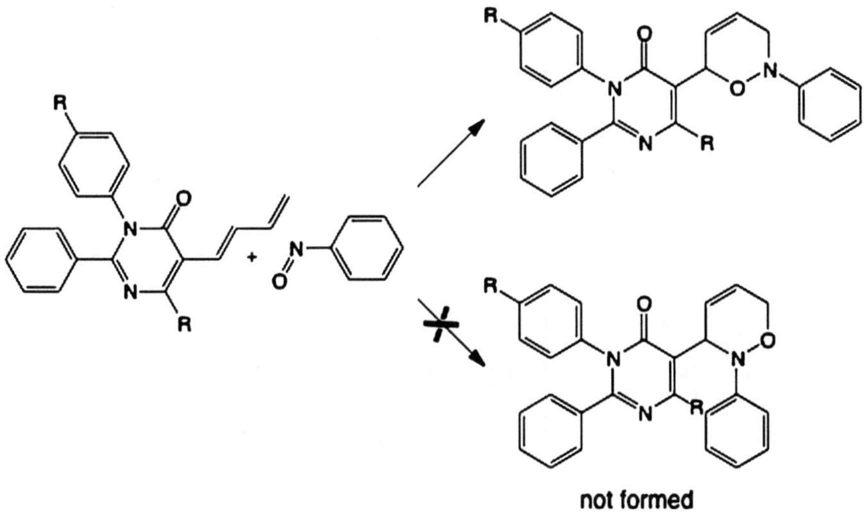

Scheme 3.50

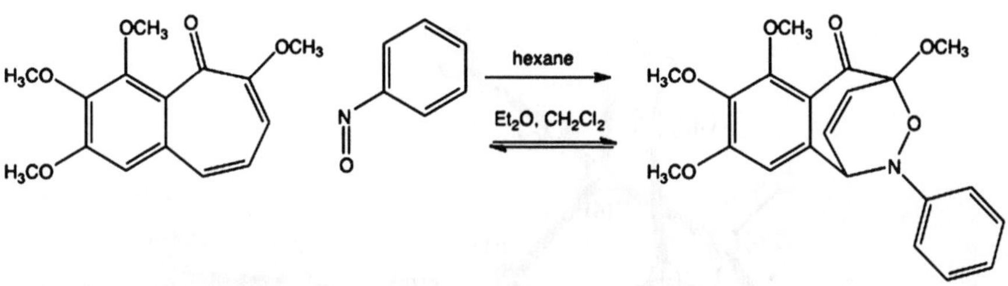

Scheme 3.51

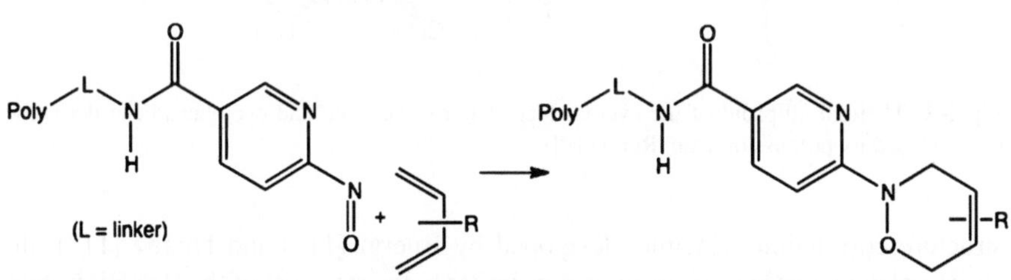

Scheme 3.52

High regioselectivity of Diels-Alder additions of nitrosobenzene could be also obtained by using pyrrolidine-based catalysts (Scheme 3.53). In combination of the Diels-Alder reaction with the tandem *O*-nitroso aldol/Michael reaction [115], only one of the isomers is obtained in the high yield. The mechanism of the reaction has been studied in details by *ab initio* calculations [116].

Since cycloadditions are reversible processes, the mechanisms of these reactions can be also studied by analyzing retro-Diels-Alder pathway. As it follows from the

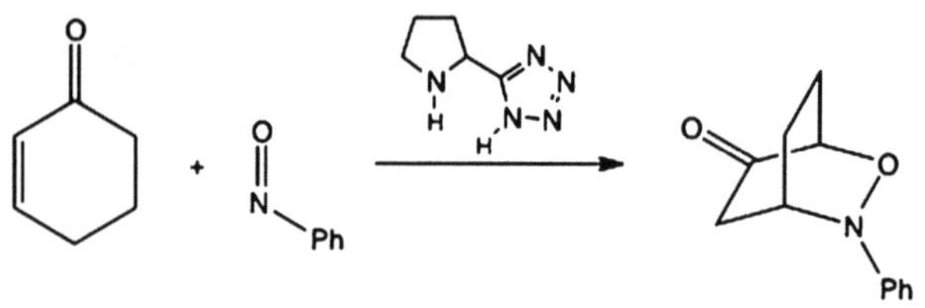

Scheme 3.53

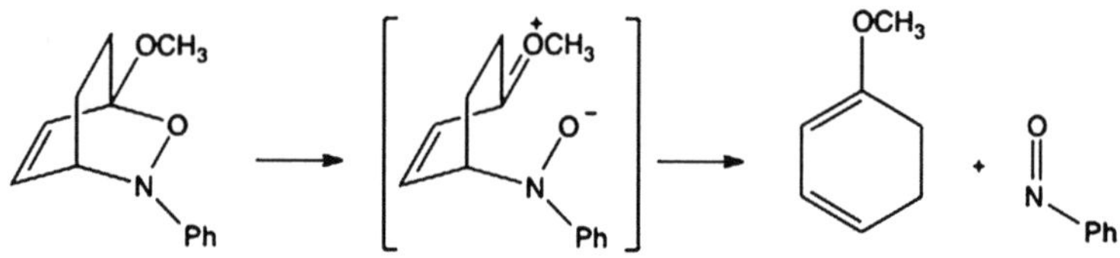

Scheme 3.54

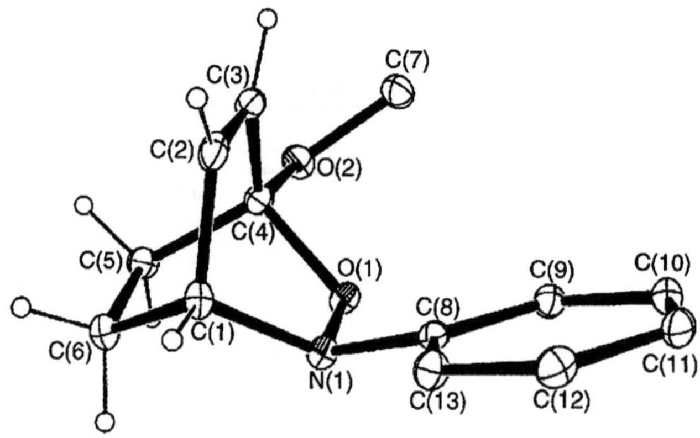

Fig. 3.2 Thermal ellipsoid of the cycloadduct of nitrosobenzene and cyclohexadiene derivative (Reproduced by permission from Ref. [119])

structure-correlation principle, developed by Buergi [117] and Dunitz [118], the structural deformations of substrate are similar to the structure of the transition state. For instance, chemical bonds that are breaking in the transition state are elongated in the reactant molecule. In the case of the retro-Diels-Alder reactions, if one of the two breaking bonds is longer relative to some standard value for the corresponding bond type, it will break first, and the pathway will be rather stepwise. In the adduct shown in the Scheme 3.54 the C-N bond is "normal", but the C-O bond is significantly lengthened (for 0.043 Å) relative to the average bond distance. It implies that the reaction path includes zwitterionic intermediate [119].

The structure of the cyclic product determined by X-ray diffraction is shown in the Fig. 3.2.

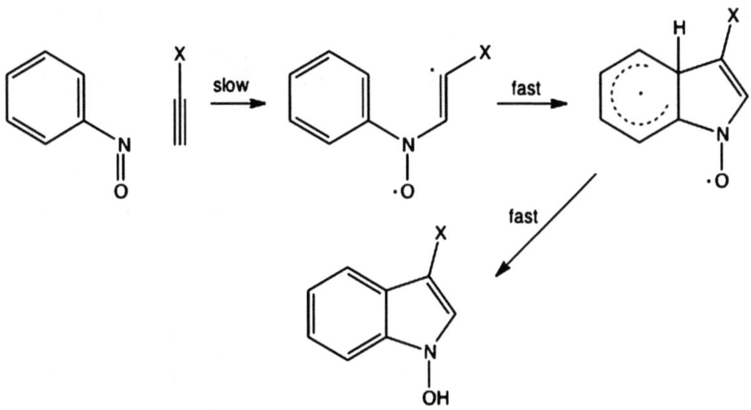

Scheme 3.55

3.1.5.1 Stepwise Cycloadditions

Formation of cyclic products by addition of nitrosoaryl compounds also operates by stepwise mechanisms. The simplest mechanism that includes biradical as a reaction intermediate was reported for a series of reactions of nitrosobenzene with alkynes. In these transformations the rate determining step is the *N*-attack of the nitrosobenzene molecule to the sterically less hindered carbon atom of the alkyne bond. The biradical intermediate (Scheme 3.55) undergoes fast cyclization to the bicyclic product [120].

The mechanism was systematically studied by using Hammett correlations as well as by measuring the kinetic isotope effects.

Biradical intermediate is also formed in cyclization which starts with nitrosobenzene and substituted alkene [121]. The mechanism of this transformation is complicated, and occurs through formation of the nitrone which by 1,3-dipolar cycloaddition with alkene yields isoxalidine derivative. The first reversible step is formation of a biradical (Scheme 3.56).

Nitrosobenzene cycloadditions also appear in multicomponent reactions (Scheme 3.57). The three-component reaction of α-diazo esters, nitrosobenzene derivatives, and electron deficient alkenes provides diverse isoxazolidines under mild reaction conditions [122].

3.2 Dimerizations and Spectroscopy

Since the nitroso-azodioxide interconversion is the process that includes molecular species (monomers and dimers), which afford markedly different spectral characteristics, the study of the dynamic and mechanism of dimerization/dissociation is

Scheme 3.56

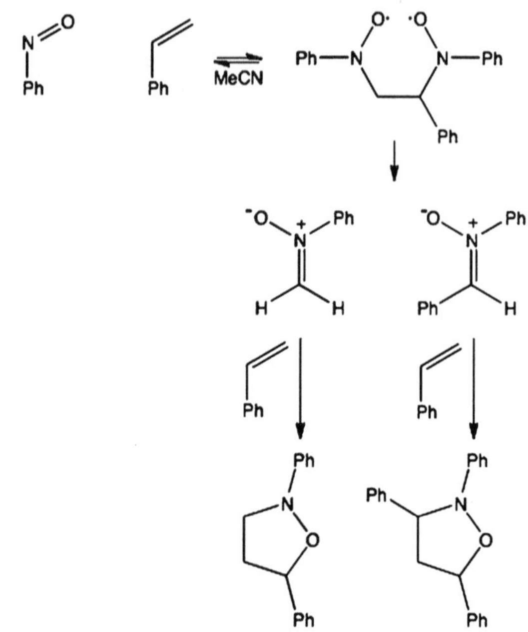

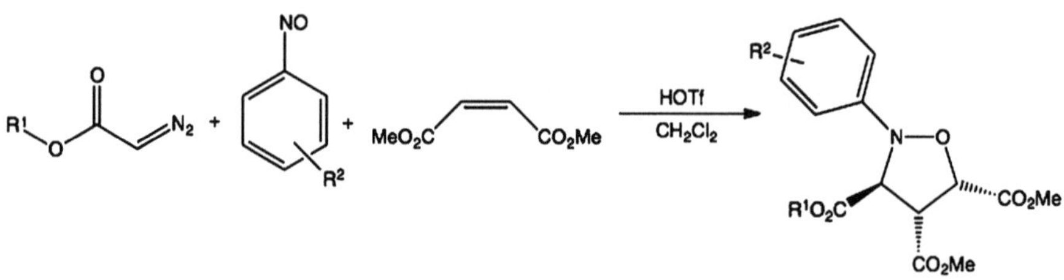

Scheme 3.57

closely connected with the spectroscopy. From that reason, the spectroscopy of nitroso compounds is discussed together with the problem of dimerization in the same chapter.

Even by visualization, it can be observed that most of the C-nitroso compounds in their crystal form appear as colorless substances, while in solution in organic solvents change the color to blue or greenish. More than 100 years ago these phenomena were elucidated by O. Piloty [123–125], who concluded, on the basis of the cryoscopic measurements of the molecular weights, that nitroso molecule in the solid-state form dimers. Fifty years later, this assumption has been confirmed by X-ray structural determinations, as well as by vibrational spectroscopy [126–129]. Crystal and molecular studies showed that nitroso molecules dimerize by forming the new nitrogen-nitrogen bond, which has a high level of double bond character [130]. The NN bondlength of 1.31 Å has an intermediate value between the single

NN bond (for instance 1.447 Å in hydrazine) and the double NN bond, which is 1.247 Å in azomethane. Formally, these dimers can be named as **azodioxides** or **diazene-1,2-dioxides**. Stereochemically, such compounds are analogous to azocompounds, so that they exist in two forms, *cis* and *trans* (*Z* and *E*). Both the stereoisomers of nitrosomethane have been for the first time prepared in the middle of the last century by B.G. Gowenlock, J. Trotman, and H.T.J. Chilton [131–133], and their structure was later confirmed by X-ray studies. Blue color of nitroso monomers has been assigned to the n-π^* transition of the n-electrons predominantly localized on the nitrogen atom. The N=N bond energy [134] (average value of 120 kJ mol^{-1}), is smaller than average covalent double bond, but almost three times larger than a hydrogen bond (20-40 kJ mol^{-1}). This property of bonding of C-nitroso dimerization, became one of the central points in investigations of this class of compounds from two aspects, mechanistic, and supramolecular, respectively. Till the last decade, several empirically equivalent hypotheses were proposed for the reaction mechanism of the dimerization. Later in this chapter, we will try to resolve the problem of the reaction mechanism in more details, primarily on the basis of recent studies of reactions in the solid-state. Formation of azodioxide bonds, on the other hand, opens the opportunity for use of dinitroso compounds as building blocks for novel type of supramolecular assemblies [135].

3.2.1 NMR Spectroscopy

In spite of the relatively simple molecular structures of nitrosoaromatic molecules their NMR spectra are rather complex, and concentration-dependent. This fact could be ascribed to the temperature-dependent mixture of monomeric, and two dimeric species. Spectra of some of relatively simple nitroso compounds, such as *o*-nitrosobenzoic acid [136], have been accurately assigned only recently. On the other side, this property of the nitroso compounds opens a new opportunity to extend the new techniques of NMR spectroscopy by using these molecules as appropriate models. As it will be demonstrated later in this chapter, some of the effects observed to have extreme values, either in the chemical shifts anisotropy or in couplings, have been found above all in the studies of nitrosoaromatic compounds.

3.2.1.1 ^{1}H NMR

Gowenlock et al. [137, 138] have published the definitive characterization of the ^{1}H NMR spectra of a series of nitrosoaryl compounds, especially substituted nitrosobenzenes. Besides the ^{1}H chemical shift data and substituent constants for shielding of several groups, which are useful for the compounds identification, the importance of the NMR spectroscopy is in demonstrating the effect of the electron donating and electron withdrawing groups on the extent of the free rotation of the

Table 3.2 ^{1}H NMR chemical shifts for the 4-substituted nitrosobenzenes

δ/ppm				*J*/Hz				
4-X	$H_{2,6}$	$H_{3,5}$	$H_{2,3}$	$H_{2,5}$	$H_{2,6}$	$H_{3,5}$	$H_{3,6}$	$H_{5,6}$
H	7.90	7.65	7.91	0.56	1.97	1.35	0.56	7.91
F	7.96	7.30	8.88	0.16	2.62	2.34	0.16	8.88
Cl	7.86	7.63	8.63	0.12	2.19	2.19	0.12	8.63
Br	7.77	7.81	8.54	0.38	2.27	2.17	0.38	8.54
I	7.60	8,03	7.91	0.06	1.35	1.97	0.06	7.91
CH_3	7.80	7.41	7.25					7.25
OCH_3	7.90	7.03	9.10					9.10

Table 3.3 ^{1}H NMR chemical shifts for the 3-substituted nitrosobenzenes

δ/ppm					*J*/Hz					
3-X	H_2	H_4	H_5	H_6	$H_{2,4}$	$H_{2,5}$	$H_{2,6}$	$H_{4,5}$	$H_{4,6}$	$H_{5,6}$
F	7.16	7.44	7.70	8.15	2.60	0	1.71	8.05	1.01	7.92
Cl	7.63	7.67	7.63	8.05	1.85	0	1.62	7.87	1.27	8.05
Br	7.77	7.84	7.57	8.10	1.91	0	1.77	7.91	1.08	7.88
I	7.99	8.11	7.42	8.03	1.73	0	1.70	7.86	1.12	7.80
CH_3	7.63	7.53	7.50	7.77	1.79	0	1.79	7.42	1.79	7.42
$CH{=}CHCO_2Et$	8.05	7.84	7.64	7.84				7.74		7.74

Table 3.4 ^{1}H NMR chemical shifts for di- and trisubstituted nitrosobenzenes

δ/ppm (measured at −105 °C)					
Substituents	H_2	H_3	H_4	H_5	H_6
3,4-Dimethyl[a]	6.08	2.11[c]	2.22[c]	7.58	9.12
3,4-Dimethyl[b]	9.14	2.42[c]	2.22[c]	7.05	6.06
3,5-Dimethyl	5.88	2.40[c]	7.37	2.40[c]	9.04
2,4,5-Trimethyl	3.25[c]	7.31	2.25[c]	2.15[c]	5.93
$3\text{-}CO_2H\text{-}4OH$[a]	9.74	12.80[d]		7.04	6.80
$3\text{-}CO_2H\text{-}4OH$[b]	6.80	12.60[d]		7.50	9.74

[a]Major rotamer
[b]Minor rotamer
[c]δ CH_3
[d]δ COOH

NO group relative to the phenyl ring. As it will be explained later in this section, free rotation is strongly correlated with the ability of nitrosobenzene derivatives to form azodioxides.

Proton chemical shifts of the selected nitrosobenzene derivatives are listed in following tables (Tables 3.2, 3.3, 3.4, 3.5, and 3.6). Hydrogen atoms are labeled as it is shown in the Scheme 3.58.

If the NMR spectra were recorded at lower temperature (−100 °C) below the rotational barrier of the NO group, the chemical shifts of all the protons become different. The largest chemical shift difference, $\Delta\delta > 3$ ppm, is found for the H_2 and

Table 3.5 ^{1}H NMR spectra of monomer heteroaromatic nitroso compounds (Ref. [168])

Compound	H_3	H_4	H_5	H_6
2-nitrosopyridine	7.40	8.12	7.72	8.82
3-methyl-2-nitrosopyridine	3.35 (CH_3)	8.11	7.54	8.36
4-methyl-2-nitrosopyridine	7.24	2.55 (CH_3)	7.52	8.65

Table 3.6 ^{1}H NMR spectra of *Z*-dimer heteroaromatic nitroso compounds (Ref. [168])

Compound	H_3	H_4	H_5	H_6
2-nitrosopyridine	7.80	7.96	7.36	8.08
3-methyl-2-nitrosopyridine	2.53 (CH_3)	7.65	7.19	7.96
4-methyl-2-nitrosopyridine	7.61	2.48 (CH_3)	7.15	7.92

Scheme 3.58

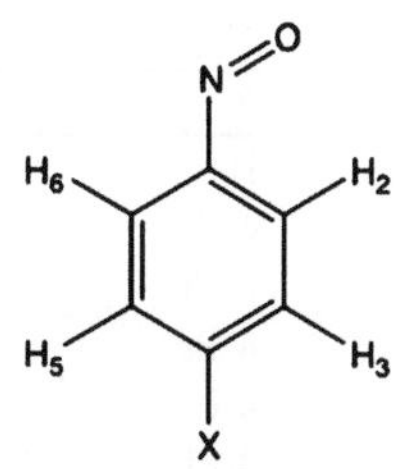

Scheme 3.59

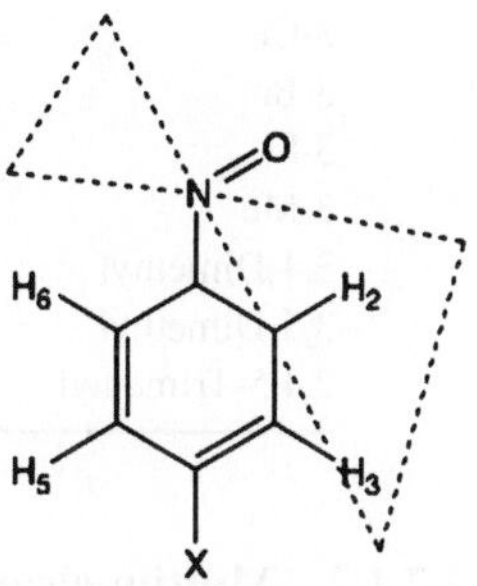

H_6 protons. Such shielding-deshielding of the protons that are *syn* or *anti* relative to the static N=O bond, represents the magnetic anisotropy of the nitroso group. In the papers written by Gowenlock's research group [137, 138] such anisotropy is described with the double cone diagram shown in the Scheme 3.59.

Interestingly, if substituent X is a strong electron donor such as methoxy or dialkylamino group, the difference in the H_2, H_6 chemical shifts, $\Delta\delta$, diminishes to 2.31, and 2.80, respectively, as it is shown in the Table 3.7.

In the dimeric forms, such H_2,H_6, and H_3,H_5 shielding-deshielding effect is negligible for all 4-substituted *E*- and *Z*-dimers perhaps because at temperatures as low as −90 °C the ring rotation is so fast that no distinct rotamers were detected. The exceptions are *ortho* substituted dimers where the rotation barrier is higher due to the steric hindrance (Table 3.8).

Table 3.7 Difference in H_2, H_6 chemical shifts for nitrosobenzene substituted with various groups in *para* position

Δδ			
X	H_2,H_6	H_3,H_5	References
H	3.24	0.55	[136]a
F	3.12	0.51	[136]a
Cl	3.20	0.57	[136]a
Br	3.21	0.44	[136]a
I	3.22	0.54	[136]a
Me	3.08	0.53	[136]a
OMe	2.80	0.42	[136]a
N(H)Me	2.34	0.25	[136]b
NMe_2	2.31	0.29	[136]b
NEt_2	2.03	0.29	[136]b

Table 3.8 ^{1}H NMR chemical shifts for Z-dimers

δ/ppm					
X	H_2	H_3	H_4	H_5	H_6
H	7.3-7.4 Multiplet				
4-F	7.32	7.03	55.0(^{19}F)	7.03	7.32
4-Cl	7.26	7.33		7.33	7.26
4-Br	7.20	7.48		7.48	7.20
4-I	7.11	7.71		7.71	7.11
4-Me	7.22	7.13		7.13	7.22
3-F	7.22	56.40(^{19}F)	7.15	7.37	7.22
3-Cl	7.55		7.42	7.32	7.16
3-Br	7.71		7.50	7.24	7.18
3-I	7.90		7.74	7.07	7.17
3-Me	7.24	2.27($C\underline{H}_3$)	7.16	7.13	6.99
3,4-Dimethyl	7.21	2.20($C\underline{H}_3$)	2.20($C\underline{H}_3$)	7.00	6.91
3,5-Dimethyl	6.93	2.21($C\underline{H}_3$)	6.96	2.21($C\underline{H}_3$)	6.93
2,4,5-Trimethyl	2.03($C\underline{H}_3$)	6.97	2.13($C\underline{H}_3$)	2.34($C\underline{H}_3$)	6.77

3.2.1.2 Multinuclear NMR

Structural investigations, especially those related to the difference between monomeric and dimeric species, can be substantially extended by measuring the chemical shifts of nuclei of atoms that are directly included in the formation of chemical bond during the dimerization. The most important information is expected from studying the chemical shift anisotropy (CSA), i.e. the directional shielding in relation to the molecular geometry.

For the starting information, the average (isotropic) multinuclear chemical shifts [139], measured for the parent nitrosobenzene are shown in the Table 3.9. Chemical shift references are Me_4Si for ^{13}C, $MeNO_2$ for ^{15}N, and D_2O for ^{17}O.

To explain these data it is necessary to investigate the details of the carbon, nitrogen and oxygen NMR spectra in the sections that follow.

Table 3.9 Multinuclear chemical shifts for nitrosobenzene

δ/ppm				
Nucleus	Label	Monomer	*Z*-Dimer	*E*-Dimer
^{13}C	C_1	165.67	143.11	142.26
^{13}C	C_2,C_6	120.99	124.42	123.63
^{13}C	C_3,C_5	129.38	129.55	128.98
^{13}C	C_4	136.00	130.99	131.32
^{15}N	N	518.7	−77.1	−80.2
^{17}O	O	~1,524	~1,524	~1,524

3.2.1.3 ^{13}C NMR

Analysis of the ^{13}C NMR spectra of nitrosoaromatic molecules is of interest not only for spectral identification, but even more for the specific properties that nitroso group affords as a substituent. We have already discussed that the role of NO group as a substituent in modifying reactivity of aromatic molecules follows from its electron accepting property (its Hammett σ_p constant is one of the highest). It has also been found that, in contrast to the nitro group, nitroso substituent acts predominantly by resonance. Since ^{13}C chemical shifts can be useful as a measure for the electron density distribution, the ^{13}C NMR spectroscopy appears to be the efficient method for the detailed study of the NO substituent effect. Later in this chapter, it will be shown that the nitroso substituent affords also unusual chemical shift pattern in *para* substituted nitrosobenzenes. In such way, NMR spectroscopy gives us additional arguments for the notion about the unique behavior of the nitroso group as a substituent.

Before developing the discussion about the special properties of the particular NO substituted aromatic molecules, let us systematize the chemical shift data for a series of substituted nitrosobenzenes. The ^{13}C chemical shifts of the nitrosoaromatic compounds, measured in solution [140–143], are listed in the following tables (Tables 3.10 and 3.11).

Analysing chemical shifts of the C_4 atom for all the listed *para*-substituted nitrosobenzenes, an interesting pattern emerges. The C_4 chemical shift is at the highest value in the case of the parent nitrosobenzene, no matter whether the *para*-substituent is a π-donor or π-acceptor. This unique behavior of the NO group has been explained by accepting the possibility that the nitroso group acts in a different way when the *para*-substituent is π-donor than when the *para*-substituent is the π-acceptor [140]. It seems that the electron demand of the nitroso group depends on the character of the *para*-substituent. As we have shown in Sect. 3.1.1, NO group exhibits an electromeric effect (σ_R, 0.33) [19], in the contrast to the nitro group, which acts by both, resonance and by strong inductive manner. Since the resonance effect is "flexible" because delocalized electrons can easily redistribute, any substituent that operates predominantly through resonance can switch its nature from electron donor to the electron acceptor.

Table 3.10 ^{13}C NMR chemical shifts (in ppm) of *p*-substituted nitrosobenzenes

X	C_1	C_2	C_3	C_4
H	135.5	129.4	120.9	166.1
Cl	142.5	129.7	122.2	163.9
Br	131.6	132.7	122.1	163.9
I	147.1	138.5	124.8	163.7
F	143.8	116.3	124.7	164.9
MeO	141.7	114.0	125.8	164.6
COMe	141.1	129.8	120.8	164.2
CO_2Me	135.3	131.0	120.3	164.6
CN	117.4	133.9	120.7	162.2
NO_2	151.2	125.5	121.3	162.7
Me	146.9	129.5	121.1	165.4
Et	151.5	128.4	115.1	165.6
i-Pr	157.4	127.1	121.3	165.8
t-Bu	159.7	126.0	121.0	165.2
NHMe	156.2	110.9		163.9
NMe_2	155.5	110.1		163.0
NHEt	154.7	110.9		164.0
NEt_2	153.4	109.8		162.9

Table 3.11 ^{13}C NMR chemical shifts (in ppm) of *o*-substituted nitrosobenzenes [144]

X	C_1	C_2	C_3	C_4	C_5	C_6
2-I	162.1	109.2	141.4	136.6	128.2	109.0
2-Me	164.9	142.1	132.9	136.1	125.6	107.3
2-OMe	162.0	158.8	115.3	138.8	119.6	108.8
2-*t*-Bu	165.7	152.2	125.8	135.4	127.7	106.5
2-Me-6- *t*-Bu	167.5	118.0	125.4	130.7	133.4	153.6
2,5-Di-*t*-Bu	165.9	148.9	127.6	132.3	149.7	103.3
2,6-Di-*t*-Bu	176.3	142.4	125.7	129.9	125.7	142.4
2,4,6-Tri-*t*-Bu[a]	174.5	141.9	122.1	151.2	122.1	141.9

[a]Solid state

3.2.1.4 ^{14}N NQR

Nuclear quadrupole resonance (NQR) spectra of the ^{14}N nucleus have been recorded and calculated for *N,N*-dimethyl-4-nitrosoaniline and the parent nitrosobenzene in solid-state. The electric field gradients, quadrupole coupling constants, χ, and asymmetry parameters η are represented in the Table 3.12 [28].

Orientations of the vector components, q, are represented in Scheme 3.60.

The largest differences between monomers and dimers are in the q_{zz} component, which represents the contribution of nitrogen n-electrons.

Table 3.12 Electric field gradients, quadrupole coupling constants, χ, and asymmetry parameters η for *N,N*-dimethyl-4-nitrosoaniline (Data from Ref. [28])

Compound	Elecric field gradient components				
	q_{zz}	q_{yy}	q_{xx}	χ	η
N,N-dimethyl-4-nitrosoaniline monomer	$(+)1.196 \times 10^{22}$	$(-)8.810 \times 10^{21}$	$(-)3.155 \times 10^{21}$	5,790	0.473
	$(+)1.198 \times 10^{22}$	$(-)9.227 \times 10^{21}$	$(-)2.761 \times 10^{21}$	5,860	0.503
a	$+1.449 \times 10^{22}$	-1.204×10^{22}	-2.455×10^{21}	7,163	0.661
Nitrosobenzene dimer	$(-)5.167 \times 10^{21}$	$(+)4.458 \times 10^{21}$	$(+)5.813 \times 10^{20}$	2,500	0.775
a	-4.951×10^{21}	$+4.308 \times 10^{21}$	$+4.848 \times 10^{20}$	2,447	0.739

[a]Labels the calculated values

Scheme 3.60

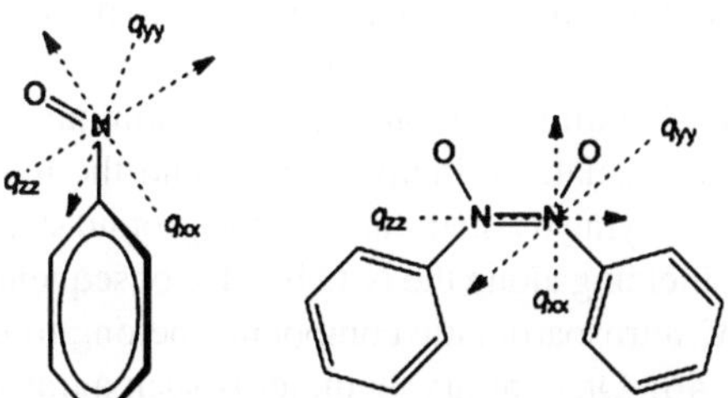

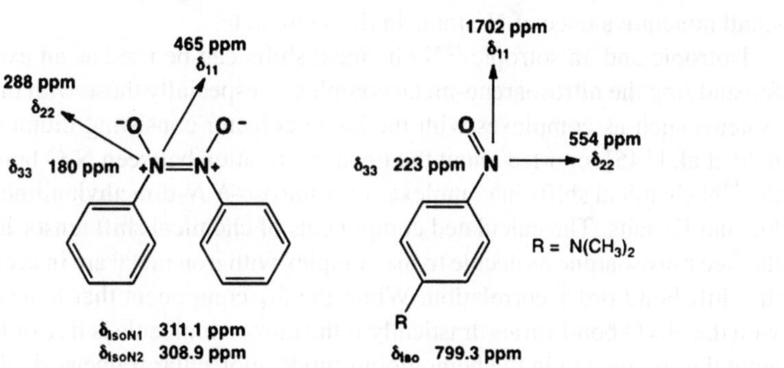

Fig. 3.3 Solid-state NMR ^{15}N chemical shift anisotropy for nitrosobenzene dimer and monomer derivative (Data from Ref. [30])

3.2.1.5 ^{15}N NMR Spectra

Variation in the molecular electron density distribution in going from the nitroso monomer to the corresponding dimer can be prominently demonstrated by analyzing the ^{15}N chemical shift anisotropy (CSA) in the solid-state NMR. The principal components of the chemical shifts measured for both forms are shown in Fig. 3.3 [30].

Scheme 3.61

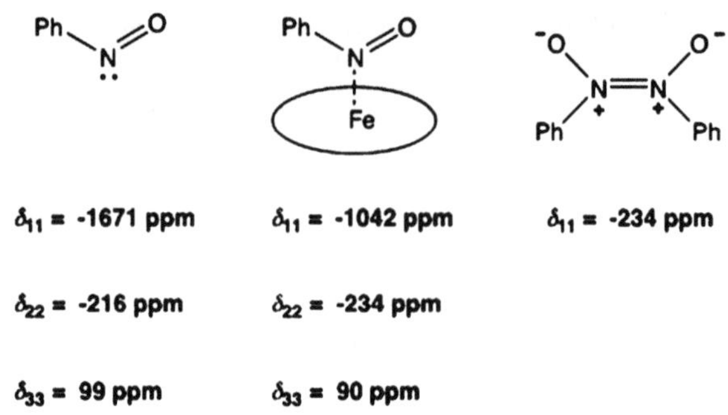

Except for the direction δ_{33}, which is perpendicular to the molecular plane, chemical shifts reveal marked differences between two molecular forms. The largest deshielding in monomer relative to dimer has been observed for δ_{11}, the component that lies relatively close to the N-O bond. It is assumed [30] that such large deshielding in monomer is predominantly due to the $n_N \rightarrow \pi^*$ transition, which have small excitation energy, and the magnetic field-induced mixing between the low-lying excited state and the ground state that results in a significant paramagnetic shielding along the N-O bond. Consequently, the ^{15}N chemical shifts anisotropies in C-nitrosoaromatic compounds belong to the largest reported in the literature [145, 146]. On contrary, in dimer is such a transition impossible because the nonbonding electrons are included in the double bond between nitrogen atoms.

Isotropic shifts for two nitrogens in dimer are slightly different indicating the small nonequivalence of N-atoms in the solid-state.

Isotropic and anisotropic ^{15}N chemical shifts can be used as an excellent probe for studying the nitrosoarene-metal complexes, especially those with the *N*-bonding systems such as complexes with the heme-cofactor consisting biomolecules. Oldfield et al. [145] reported about the linear correlation between N-O bond orders and the ^{15}N chemical shifts in complexes of *p*-nitroso-*N,N*-dimethylaniline with Zn, Pd, Sn, and Fe salts. The calculated components of chemical shift tensor in going from the free nitrosoarene molecule to the complex with iron metal are in accordance with the shift-bond order correlation. While the δ_{11} component that is nearly collinear with the N=O bond varies drastically if the nitroso molecule is free or bound to iron atom that is situated in the hemoglobin-model molecular framework, the remaining components, δ_{22} and, δ_{33} are in principle invariant. If the nitrosoarene molecule forms dimer, the shift is normally smaller because no n_N-electron contribution is available (Scheme 3.61).

Conceptually, the shielding tensor could be divided in three regions, A, B, and C (Scheme 3.62) [145]. The dominant influence on the shielding field is the region A where the lone pair is situated, and which has the main effect on the δ_{11} component.

Scheme 3.62

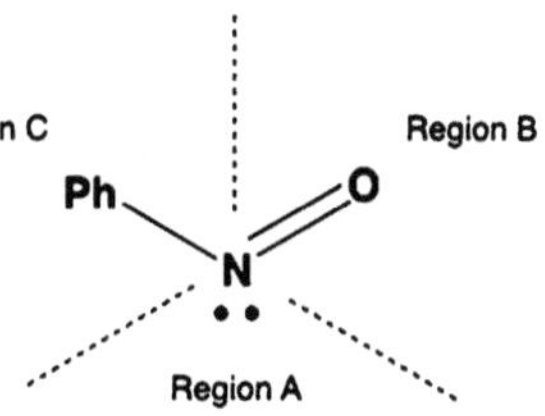

Scheme 3.63

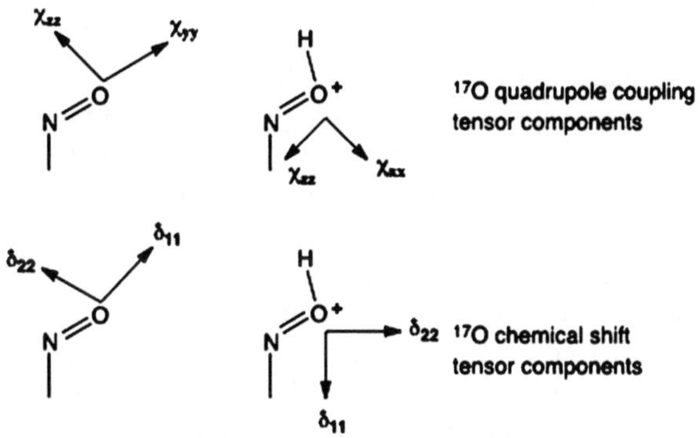

3.2.1.6 ^{17}O NMR

Investigations of the ^{17}O NMR spectroscopy of nitrosoarenes are important from two reasons. First, as in the case of the ^{15}N spectroscopy, the ^{17}O chemical shifts and chemical shifts anisotropy of the nitrosoaromatic species are extremely large [139, 147] and this observation could afford new insights into the nature of these molecules. Secondly, from the same reasons the use of nitrosoarenes in these studies can be used as a convenient test system for developing the ^{17}O NMR spectroscopy more generally.

Wu et al. have studied the anisotropy in the differently complexed nitrosoaromatic molecules by measuring both, the chemical shift and the quadrupole coupling tensors [148]. They have found that both the tensors are sensitive to the mode how the nitroso group is bound to the metal center. The isotropic ^{17}O shift changes from 1,300 ppm in *N*-coordinated complexes to 1,000 ppm in the *O*-coordinated complexes. The similar trend also reveals values of quadrupole coupling constants, which are in *N*-complexes (13 MHz) larger than in the *O*-complexes (11 MHz). Orientations of components of both tensors are shown in Scheme 3.63. To remind, δ_{33} is oriented perpendicularly to the molecular plane. In pure nitroso molecule the χ_{yy} is collinear with the N=O bond, and the δ_{11} and the N=O bond line are under the angle of 7-16°. However, in protonated form the nitroso group bond-line is under 20° to the χ_{zz} component, and under 70° to the δ_{11} component.

Similarly to the ^{15}N NMR spectral data, the chemical bond orders are in linear correlation with the components of chemical shift tensor. More precisely, the three components are differently correlated to the π-bond contribution of the bond orders. As it could be expected from the discussion in the previous section, the largest bond-order dependence is found for δ_{11}, while δ_{33} affords no correlation.

3.2.2 UV-VIS Spectroscopy

Aromatic nitroso compounds dissolved in organic solvents show three maxima in electronic spectra, two at nearly 280 and 305 nm, and one in the visible region at nearly 730 nm. While by lowering the temperature, the signal at 730 nm diminishes, the bands in the UV region grow in the intensity. The intensity ratio of two UV signals is independent on concentration, but changes with the polarity of the solvent [39]. The shortest wavelength signal (280 nm) is assigned to the π-π^* transition, and the maximum in the 730 nm visible region to the n_N-π^* transition [149]. The absorbance at 305 nm belongs to the n_O-π^* excitation [150]. Such assignation is confirmed by findings that by *ortho* substitution with methoxy group, the UV maxima are shifted to the longer wavelengths, while the visible maximum is shifted to the shorter wavelength. The red shift of the UV signals can be explained by the delocalization of π electrons through the benzene ring. The blue shift of the signal in the visible 730 nm region is more pronounced if the *p*-substituents are stronger electron donors. In *p*-nitrosophenol the n_N-π^* transition appears at 697 nm, what is for 58 nm shorter wavelength then in the parent nitrosobenzene. Such a spectral behavior clearly shows that electron donating groups strongly stabilize the quinoide-like structure. Oppositely, the electron withdrawing substituents cause the red shift. In the spectrum of *p*-nitronitrosobenzene, the n_N-π^* signal appears at 780 nm [151]. Spectral positions of all the maxima are also sensitive on the polarity of the solvents. In polar solvents the π-π^* transition is red shifted, and the n_N-π^* transition is blue shifted.

On the basis of the recently reported high level quantum chemical calculations ((PCM-)TD-PBE0/6-311++G(3d,3p))//((PCM-)TD-PBE0/6-311G(d,p)), the n_N-π* excitations in different nitroso compounds were interpreted as HOMO-LUMO transition [152]. While the HOMO orbital possesses a significant *n*-character with the contribution of the C-N bond, the LUMO orbital has a nodal plane that intersects this bond. These findings can clarify the ability of the C-nitroso molecules to photodissociate in two radicals, NO and the rest of the molecule.

Nature of the UV-VIS absorption spectra of the aromatic nitroso compounds also depends on the structural rigidity of the molecule. Zharkova et al. have calculated dipole moments and UV-VIS absorption spectra of 1-nitroso-2-naphthol, and they have found that the best agreement between the calculated and the recorded spectra is for the structure with the intramolecular hydrogen bond [153].

Visible and near UV absorptions were analysed in detail from the spectra of nitrosobenzene recorded in argon matrices at 12 K [154]. Since the spectra afforded

Table 3.13 Spectral properties of substituted nitrosobenzene derivatives (Data from Ref. [155])

Substance	λ_{max}/nm
Nitrosobenzene	754.7
4-bromonitrosobenzene	757.5
2-iodonitrosobenzene	776.7
4-iodinitrosobenzene	757.5
4-nitroso-N,N-dimethylanilin	719.4
2,4,6-tribromonitrosobenzene	775.2
Nitrosomesitylen	796.8

vibronic structure, the vibrational frequencies of the ground state, S_1, as well as for the two excited states (S_1 and S_2) were observed and partially assigned. The first UV band at 30,260 cm^{-1} was assigned to the $S_0 \rightarrow S_2$ transition, and, on the basis of the line shape analysis, it was estimated that the S_2 excited state has lifetime nearly 80 fs. Additionally, it was found that the excess of excitation energy of 2,200 cm^{-1} causes fragmentation into phenyl radical and NO. The importance of this discovery will be explained later, when the photodissociation of nitrosobenzene will be discussed.

Some of the characteristic absorption maxima assigned to the $n < -\pi^*$ transition in the substituted nitrosobenzene monomers recorded in the benzene solution are represented in the Table 3.13 [155].

3.2.3 Photoelectron Spectroscopy

Most of the published works about the photoelectron spectra of nitroso compounds, especially aromatic, deal with the X-ray spectroscopy (XPS or ESCA). Since the shift in the core electron binding energy (E_B) correlates with the atomic charges [156], X-ray photoelectron spectroscopy can be helpful in studying the monomer-dimer equilibrium, because the atomic charges on nitrogen and oxygen atoms markedly differ in the free nitroso relative to the azodioxide forms. The binding energies, E_B, measured for a series of nitrosoaromatic compounds are listed in the Table 3.14 [157].

From the represented data for E_B, the shift for the 1s electron from the N-atom is nearly for 3.4 eV larger in dimers than in monomers. On the other side, the shift for the oxygen 1s electron differs only slightly (2 eV) and in opposite direction. The effect was explained as a transfer of electron density from nitrogen to oxygen upon going from monomer to dimer form.

XPS spectra of nitrosobenzene are also useful for the study of the surface chemistry of monomolecular layers. On the way to the development of molecular electronics based on the lateral conductance of electrons via aromatic π-electrons, the cyclo-adduct of nitrosobenzene molecule on the Si(000) surface has been studied. It was found that the 1s binding energy of the N-atom falls in the 399.8-400.08 eV range. Since the same spectrum has been obtained with the

Table 3.14 ESCA core binding energies for O and N atoms of the nitroso groups in nitrososbenzene derivatives (Data from Ref. [157])

	Core electron binding energies in eV		
	E_B(N)	E_B(O)	ΔE_B(O-N)
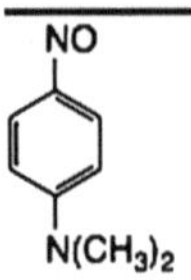	400.3	534.1	133.8
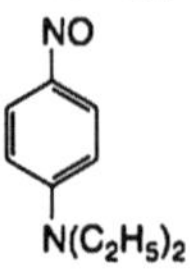	399.7	533.1	133.4
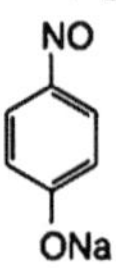	400.1	532	132
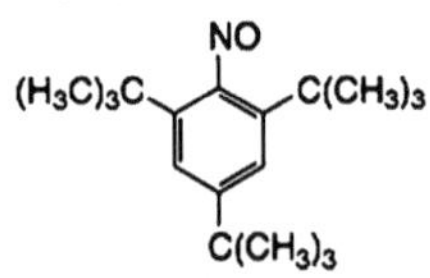	401.6	533	132
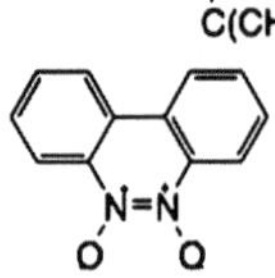	403.8	531.8	128
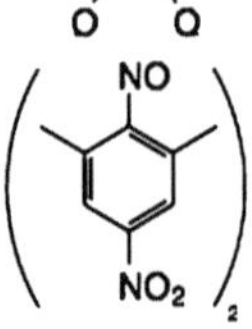	403.7	532.5	128.8

deposition of nitrobenzene, the reaction on the silicon surface is explained as the reduction of nitrobenzene to the nitrosobenzene, and finally to the nitrene adduct (Scheme 3.64) [158, 159].

The ultraviolet photoelectron spectroscopy (HeI, and HeII) of nitroso compounds [160] has been less investigated because the monomer-dimer equilibrium complicates interpretation of the spectra. However, in the early experimental works, the HeI spectroscopy has been applied for studying this equilibrium in the gas phase [161–163].

In the spectrum of the parent nitroso-compound, nitrosomethane, the three distinct bands have been resolved [163]. First band consists from the ionization potentials (IP) at 9.76 and 13.8 eV, which were assigned to the ionizations from the n-orbitals. The second broader system at 14.3 eV origins from the N=O π-orbitals, and the third band with peaks at 15.8 and 16.9 eV was interpreted as the ionization from the pseudo-π orbital of the methyl group. The details of the spectral assignation are obtained on the basis of the high-level quantum chemical calculations (MP2 and DFT) and are listed in the Table 3.15 [164].

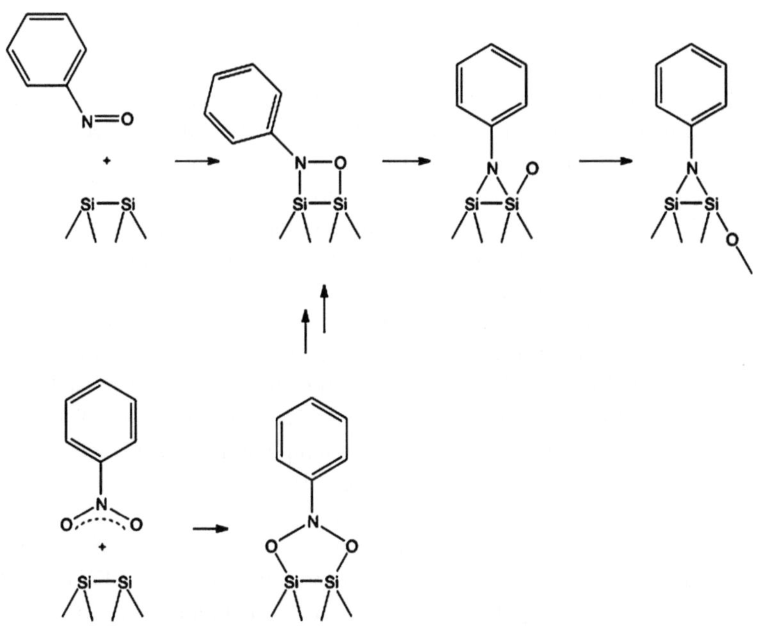

Scheme 3.64

Table 3.15 Assignation of photoelectron spectrum of nitrosomethane (Data from Ref. [164])

Assignation	Orbital	Experimental IP/eV	Calculated IP/eV
$10a'^{-1}$	σ_{NC},$p_{y(O)}$	9.76	9.28
$9a'^{-1}$	$\pi_{y(CH3)}$	13.8	13.38
$2a''^{-1}$	$\pi_{z(NO)}$-$\pi_{z(CH3)}$	14.3	14.9
$1a'^{-1}$	$\pi_{z(NO)}$-$\pi_{z(CH3)}$	15.8	15.36
$8a'^{-1}$	σ_{NO}	16.9	16.23

Unfortunately, the photoelectron spectrum of nitrosobenzene is difficult to interpret, because the spectral region of the N=O π-orbitals is covered by signals of the benzene π-system [165]. First ionization potential of *p*-substituted nitrosobenzenes, assigned to the ionization of *n*-electrons, strongly depends on the nature of the substituent [163]. While in the parent nitrosobenzene this energy is 8.51 eV [165] (the value 8.90 eV has also been recorded) [166] in 4-nitroso-*N*,*N*-dimethylaniline it is much lower, 7.78 eV. Chlorine atom in *para* position affords an opposite effect, the 4-chloronitrosobenzene has its first ionization potential at 9.02 eV [162].

Recently, a threshold photoelectron spectrum (TPES) of nitrosobenzene has been recorded and interpreted [167]. The spectral data were used for determination of the heat of formation of nitrosobenzene, which is 198.6 kJ mol^{-1} at 298.15 K.

3.3 Monomer-Dimer Equilibrium

After the first discovery of geometrical isomers by Gowenlock et al. [131–133], monomer-dimer equilibrium of various nitroso compounds has been intensively studied by UV-VIS, NMR, as well as by IR spectroscopy methods. In principle, the *E*- and *Z*-dimeric forms can be interconverted *via* monomer species in solution (Scheme 3.65). The dimeric forms are predominant at temperatures lower than −30 °C. However, there are exceptions such as 2-nitrosopyridine, which appears as *Z*-dimer at the room temperature [168].

In the crystal state, most of the substituted nitrosobenzenes appear in the *E*-form, with the exception of the parent nitrosobenzene, 3-fluoronitrosobenzene, 3-methylnitrosobenzene, and 3,5-dimethylnitrosobenzene, which crystallize in the *Z*-configuration. Some of derivatives after a slow evaporation of the solvent crystallize as pure *Z*-isomer, but at the room temperature they convert to the more stable *E*-isomer. The following table (Table 3.16) lists the appearance of various nitrosobenzene derivatives in the crystal form.

First systematic study of such equilibrium in solution has been made for aliphatic nitroso compounds by using UV spectrometry data [173–175]. For *Z*- and *E*-azodioxymethanes and cyclohexanes it was found that monomers dimerize at low temperature to the *Z*-isomer by kinetic control, whereas at higher temperature these compounds dimerize to the *E*-isomer by thermodynamic control.

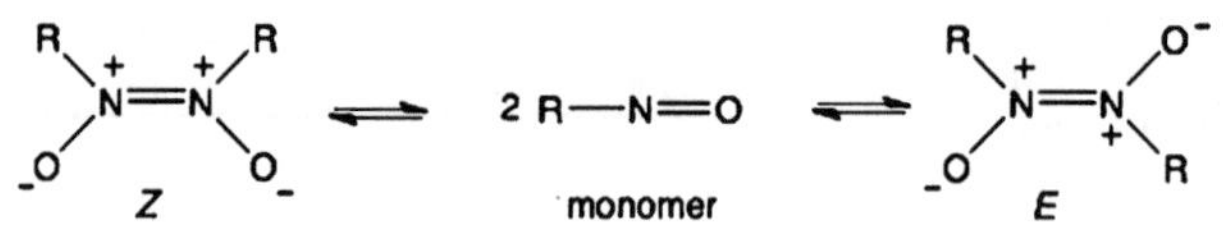

Scheme 3.65

Table 3.16 Solid-state structures of substituted nitrosobenzenes

Substituent	Isomer	References
4-F	*E*-dimer	[169]
4-Cl	*E*-dimer	[169]
4-Br	*E*-dimer, monomer	[169–171]
4-I	*E*-dimer, monomer	[26, 172]
4-Me	*E*-dimer	[169]
4-OMe	*E*-dimer monomer	[8, 172–176]
3-F	*Z*-dimer	[177]
3-Cl	*E*-dimer, *Z*-dimer	[177]
3-Br	*E*-dimer, *Z*-dimer	[177]
3-I	*E*-dimer, *Z*-dimer	[139, 177]
3-Me	*Z*-dimer	[177]
3-CH=CHCO$_2$Et	*E*-dimer, monomer	[177, 178]
3,4-diMe	Monomer	[177]
3,5-diMe	*Z*-dimer	[177]
2,4,5-triMe	*E*-dimer, monomer	[8, 177]
3-CO$_2$H-4-OH	Monomer	[179]

Table 3.17 Thermodynamic parameters for the nitrosobenzene monomer-dimer equilibrium

Equilibrium	ΔH (kJ mol^{-1})	ΔS (kJ mol^{-1})	ΔG (kJ mol^{-1})
Z-to-monomer	55.4	213.7	−8.1
E-to-monomer	42.5	179	−10.9

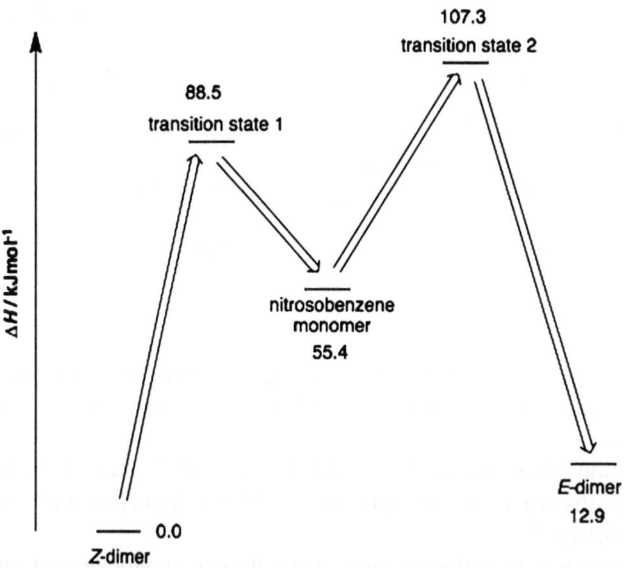

Fig. 3.4 Enthalpy diagram for dimerization of nitrosobenzene molecules

In most cases, the equilibrium is shifted to the Z-form because of its greater stability. The detailed introspection in this system in the case of the parent nitrosobenzene has been obtained by multinuclear two-dimensional NMR spectroscopy [139]. At room temperature, the spectrum of nitrosobenzene in $CDCl_3$ solution afforded only the signals of the nitroso monomer. By lowering the temperature to −30 °C these monomer peaks almost disappeared with the appearance of new signals assigned to *cis* and *trans* dimers. Thermodynamic parameters for both equilibria are shown in Table 3.17.

The most pronounced difference between *cis*-to-monomer and *trans*-to-monomer equilibrium is found in the standard reaction entropy. Larger entropy for *cis*-to-monomer is attributed to the greater partition function caused by relaxation of the restricted rotation of benzene rings. Since the restricted rotation is temperature sensitive, the larger entropy in *cis*-to-monomer causes that the Gibbs energy also vary significantly with the temperature [139]. While at 298.15 K the ΔG value is -8.1 kJ mol^{-1}, its value at 233 K is 5.8 kJ mol^{-1} [139].

From the variable temperature measurements by 2D-EXSY NMR, the thermodynamic and kinetic parameters for the dissociation reactions were calculated, and the enthalpy profile for the equilibrium, shown in Fig. 3.4, has been obtained.

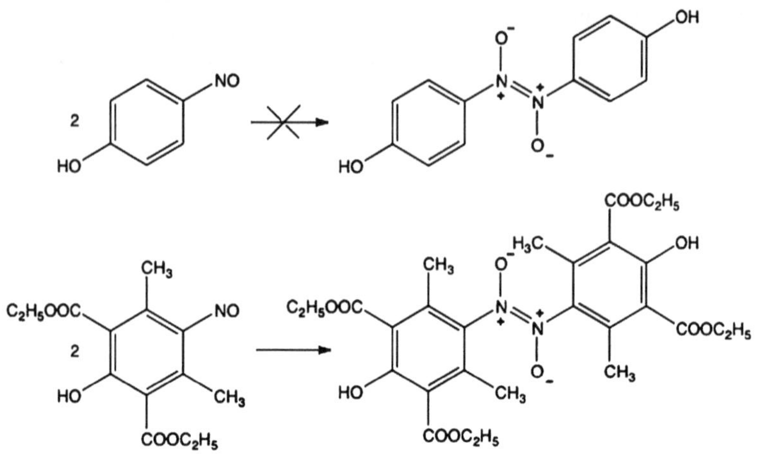

Scheme 3.66

Not only that the Z-azodioxide has lower enthalpy, but also the activation energy for dissociation to the E-isomer is for 18.8 kJ mol^{-1} higher than for formation of the Z-isomer.

The reaction rate constants (k/s^{-1}, 20 °C) for both isomers differ significantly, and while for dissociation of *cis* dimer $k = 9.63$, for dissociation of *trans* dimer its value is $k = 1.91\ s^{-1}$.

Factors which have influence on the stability of monomer and, consequently on the equilibrium, are stereoelectronic in nature [180, 181]. Generally, nitroso monomer is more reactive if the lone electron pair is more localized on the nitrogen atom. Consequently, degree of delocalization of the nitrogen n-electrons is correlated with the ability of monomer molecule to form azodioxide. Delocalization is pronounced if benzene ring and the nitroso group are coplanar. Coplanarity is enhanced by electron-donating substituents, but hindered by voluminous groups bound to the *ortho* position. Substituents in the o-position mostly afford sterical disturbance that cause twisting of the benzene ring relative to the C-N=O plane. In the twisted form, the π-delocalization of electrons from the nitroso group to benzene ring is prevented, and the contribution of the quinoide resonance structure is minimized (Scheme 3.7).

On the other hand, p-substituents strongly influence on the extent of quinoide structure, and on the dimerization. Strong electron donors, such as dialkylamino or methoxy groups promote delocalization, so that p-dimethylaminonitrosobenzene or p-methoxynitrosobenzene does not form dimers.

Demonstration of both the effects, electron-donating and sterical hindering, respectively, is formation of dimers of substituted p-nitrosophenols (Scheme 3.66). While no dimerization is observed for nitrosophenol [182] in which n-electrons from p-hydroxy group promote quinoide structure, the o-dimethylsubstituted derivative strongly forms dimers because of the additional effect of *ortho* sterical hindering and twisting of the benzene ring relative to the C-N=O (or ON=NO) plane [183].

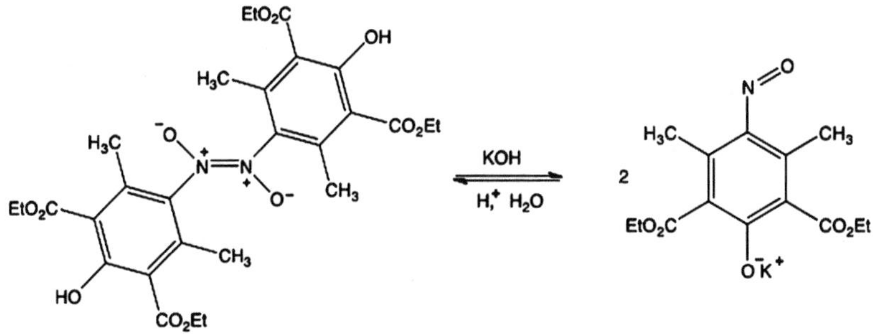

Scheme 3.67

The effect of the quinoide structure on the stabilization of the monomer form is nicely demonstrated in the systems where the electron-donating ability of the *p*-hydroxyl group in the substituted nitrosophenol is moderated by neutralization with strong base such as KOH. As it is shown in the Scheme 3.67, enhancement of the electron-donating effect of the *p*-substituent by transforming OH group to the phenolate salt leads to the formation of the stable monomer [184].

As it is demonstrated, in most derivatives the *Z*-dimer is thermodynamically more stable in solution. However, if the benzene ring is substituted in *ortho* positions, as for instance in 2,2′,6,6′-tetramethylazodioxybenzene, the repulsion between benzene rings is more pronounced in the *Z*-form, and the *E*-isomer becomes more favorable [173]. Electronic delocalization of benzene π-electrons and electrons in the ON=NO bridge is restricted. This follows from the comparison of the observed geometrical parameters of aliphatic and aromatic azodioxides, which show that the nitrogen-nitrogen bond length is almost the same in both classes of the nitroso dimers [185]. As in the case of nitrosoaromatic monomers, the rotation of the benzene ring relative to the ON=NO bridge in azodioxides is relatively free at room temperature, and becomes restricted only by lowering the temperature [186].

In contrast to the nitrosobenzene molecules, nitrosoaromatic compounds such as 2-nitrosopyridines form significant quantity of dimeric forms in solution already at room temperature. The *Z*-dimer-to-monomer equilibrium constant measured by NMR in $(CDCl_2)_2$ at 303 K is 0.208 [168].

3.3.1 Dimerizations and Polymerizations of Dinitroso Compounds

The basic idea of polymers with alternating azodioxy bonds and molecular spacers is very old and it has been published almost 130 years ago [187, 188]. Benzene ring with two nitroso groups in *m*- or in *p*-positions represents an appropriate monomeric unit for the formation of oligomeric and polymeric aggregates. In principle, two structural motifs are possible, depending whether the monomer

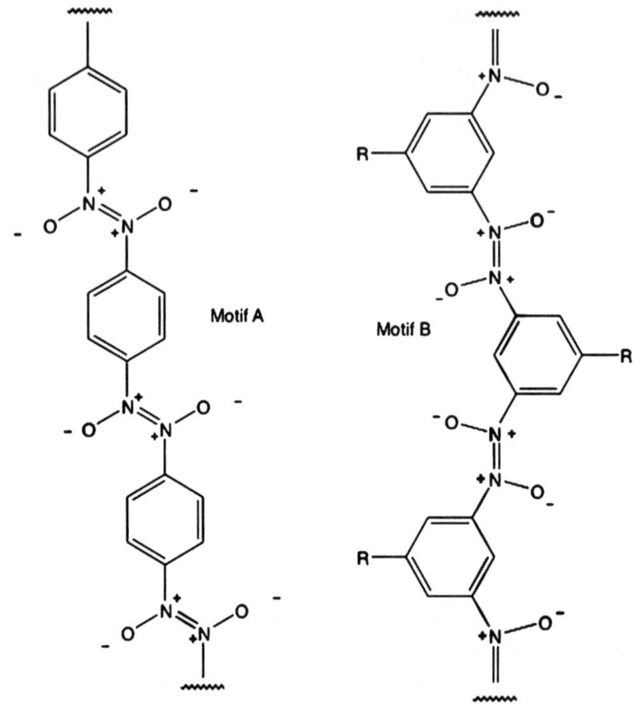

Scheme 3.68

is *m*- or *p*-dinitrosobenzene. Structures formed from *meta* monomer [189] (Motif B in Scheme 3.68) could be interesting for the architecture with interesting supramolecular arrangements, which could be designed by introducing various substituents R. On the other hand, the *p*-derivate can form polymer with closer packing of supermolecular chains (Motif A in Scheme 3.68). Variation of possible structures is even much wider if *cis*-stereochemistry of azodioxy groups is also taken into account. However, the vibrational spectroscopy [169, 190, 191] and solid-state NMR investigations [192, 193] suggests that only the *trans* configuration was present in polymers.

As it was mentioned in the Chap. 2 of this book, *p*-dinitrosobenzene appears as an amorphous polymeric solid. The monomeric form can be obtained either by heating in xylene or by sublimation to the surface cooled by liquid nitrogen [194]. From the detailed spectroscopic study of the polymerization of *p*-dinitrosobenzene isolated in argon matrix, Hacker et al. [195] have found that the polymerization appear suddenly and very fast when the film of monomers is heated to 170 K. The same temperature of dimerization was also observed for the monomers obtained by photolysis in the crystal (see Sect. 3.3.3). Such a behavior is in agreement with the reported low activation energy (ca. 84 kJ mol^{-1}) for the dissociation of the nitrosobenzene dimer [175, 196].

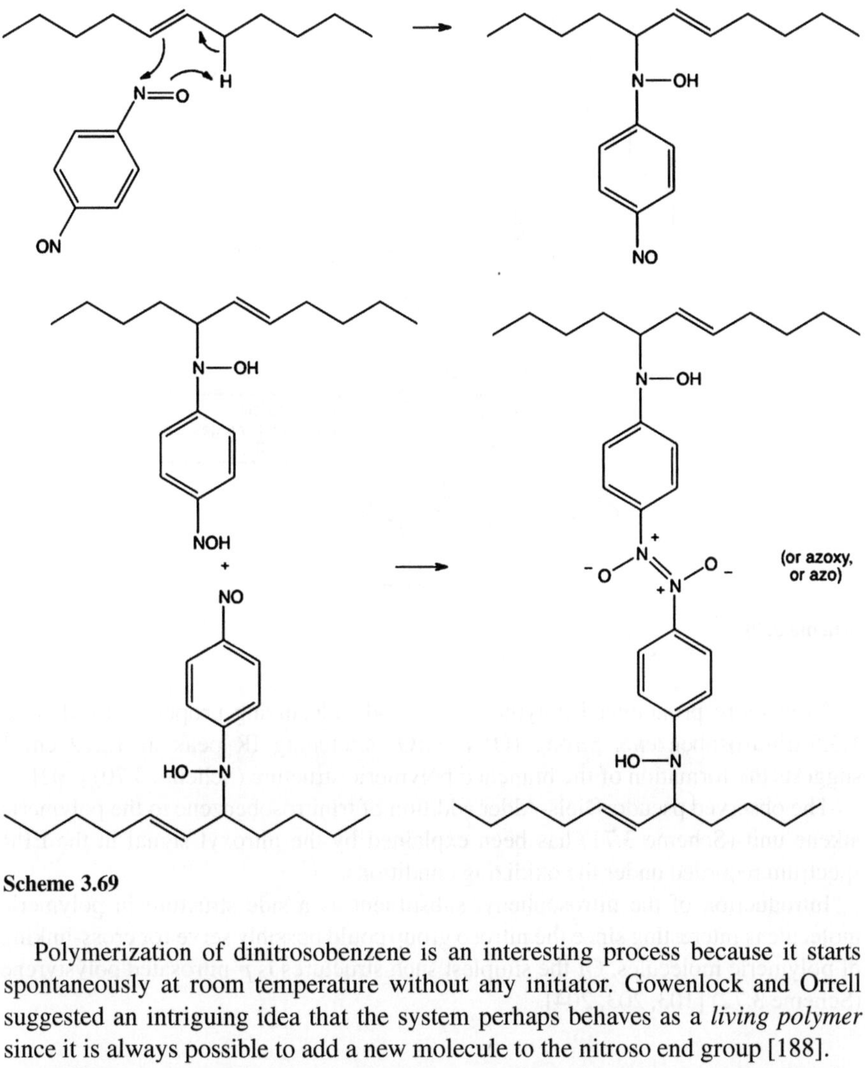

Scheme 3.69

Polymerization of dinitrosobenzene is an interesting process because it starts spontaneously at room temperature without any initiator. Gowenlock and Orrell suggested an intriguing idea that the system perhaps behaves as a *living polymer* since it is always possible to add a new molecule to the nitroso end group [188].

Additional extension of the chemistry of nitrosobenzene polymers is also possible by starting with the monomers designed with different substituents on the benzene ring. Steric effects of the groups in *ortho* position to the nitroso function could form polymers in which benzene rings are fixed at different angles relatively to the ONNO plane [197, 198].

Property of the nitroso group to form azodioxides, azoxides, and azo-bridges makes nitroso- and dinitroso-compounds convenient substances for the formation of cross-linked polymers [199]. Vulcanization of long chained allylic hydrocarbon molecules by 1,4-dinitrosobenzene has been suggested [200, 201]. Besides the reaction of the nitroso group with the allyl moiety, the cross-linking by the formation of azodioxide group (or their reduced forms such as azoxy and azo-group) has been observed (Scheme 3.69).

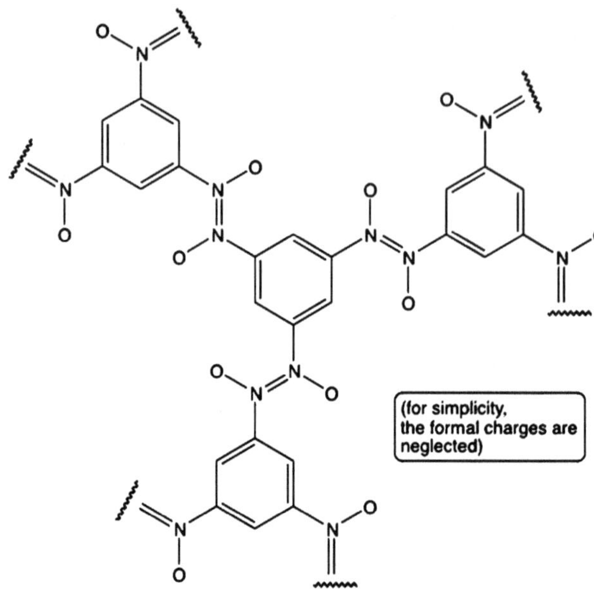

Scheme 3.70

Even more pronounced polymerization and vulcanizing property affords also 1,3,5-trinitrosobenzene. Strong (O)N=N(O) stretching IR peak at 1,272 cm^{-1} suggests the formation of the branched polymeric structure (Scheme 3.70) [202].

The observed pseudo-Diels-Alder addition of trinitrosobenzene to the polymeric alkene unit (Scheme 3.71) has been explained by the nitroxyl signal in the EPR spectrum recorded under the oxidizing conditions.

Introduction of the nitrosophenyl substituent as a side structure in polymeric molecule is interesting since the nitroso group could possibly serve for cross-linking of polymeric molecules. Of the simplest such structures is *p*-nitrosated polystyrene (Scheme 3.72) [103, 203, 204].

3.3.2 Vibrational Spectroscopy

For analysis and distinction of monomeric and dimeric structures of aromatic nitroso compounds, the most useful spectral characteristics are infrared active vibrations of monomers which include N=O stretching in the 1,488-1,513 cm^{-1} spectral region, and C-N stretching vibration in the 760-850 cm^{-1} range [8, 169, 170, 177]. While the ON=NO asymmetric stretching vibration in the 1,253-1,299 cm^{-1} range is assigned to *E*-dimers, *Z*-dimers can be recognized with absorptions in two spectral regions, 1,389-1,397 cm^{-1}, and 1,409 cm^{-1}, respectively.

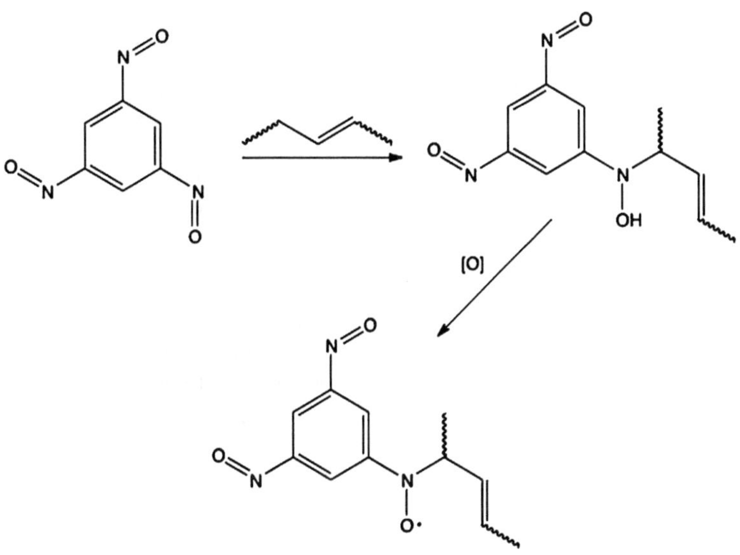

Scheme 3.71

Scheme 3.72

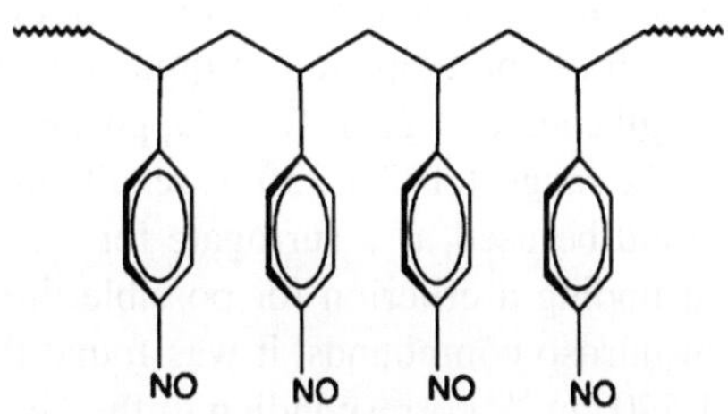

Table 3.18 IR and Raman frequencies (in cm^{-1}) of nitrosobenzene derivatives (Data from Ref. [104])

4-Methylnitrosobenzene		4-Bromonitrosobenzene		4-Chloronitrosobenzene	
IR	Raman	IR	Raman	IR	Raman
1,600 (30)	1,600 (95)	1,576 (13)	1,574 (90)	1,582 (20)	1,582 (45)
	1,436 (90)		1,442 (80)		1,444 (60)
	1,295 (95)		1,292 (90)		1,290 (70)
1,257 (260)		1,260 (95)		1,265 (140)	
	803 (100)	809 (45)		820 (85)	
760 (82)			759 (100)		766 (100)

Because of the symmetry selection rules, the intensity of the ON=NO asymmetric stretching vibration in *Z*-dimers is in the Raman spectra very low or inactive. The IR and Raman frequencies and the relative intensities of signals (in parenthesis) for the selected nitrosoaromatic dimers are represented in the Table 3.18 [104].

Examples of N=O stretching frequencies for substituted nitrosobenzene monomers are listed in the Table 3.19.

Table 3.19 N=O stretching wavenumbers (in cm^{-1}) [169, 172, 205]

Compound	$\nu_{N=O}$, cm^{-1}
Nitrosobenzene	1,506
p-Fouoronitrosobenzene	1,511
p-Chloronitrosobenzene	1,500
p-Bromonitrosobenzene	1,497
p-Iodonitrosobenzene	1,488
p-Methylnitrosobenzene	1,508
p-Methoxynitrosobenzene	1,497
p-Nitronitrosobenzene	1,513
m-Nitronitrosobenzene	1,504
m-Nitrosonitrosobenzene	1,511
3,5-Dichloronitrosobenzene	1,502
o-Methylnitrosobenzene	1,499
o-Nitronitrosobenzene	1,511
o-Me(i)thoxynitrosobenzene	1,495
o-Iodonitrosobenzene	1,502
2,4,6-Tribromonitrosobenzene	1,506

Interesting correlation between N=O stretching frequencies, N=O bondlengths, and ability to dimerize has been found by Cameron et al. [206]. In molecules in which the participation of quinoide structure is pronounced, the N=O distance is lengthened to 1.25-1.37 Å. Oppositely, molecules with shorter N=O bond lengths, in the range 1.192-1.230 Å, readily dimerize. Since the N=O stretching frequency could be used as a surrogate for the bond distance measuring its value can help in finding a criterion for possible dimerization. As a result of analyzing a series of nitroso compounds, it was found that there is a "dimerization window" 1,480-1,590 cm^{-1}, corresponding to the N=O bondlength range 1.192-1.230 Å. Nitroso monomer that has N=O stretching frequency within this range will readily dimerize. For instance, *p*-nitrosoaniline with the value of 1,363 cm^{-1} falls out of dimerizing window, and does not dimerize at all. It is important to point out that there is no correlation between N=N bond length of dimers and the ease of dissociation [207]. Therefore, the ability for dimerization is strongly dependent on the properties of monomer. This conclusion will be important in our discussion of the reaction mechanism of dimerization, later in this book.

3.3.3 *Kinetics of Dimerization in Solid-State*

Previously we have discussed the solid-state dimerization and polymerization in the amorphous solids. However, for investigation of dimerizations in the crystal or polycrystal state, it is necessary to provide crystals in which monomer is enough stable, and to find conditions under which the dimerization reaction halftime falls in the range where it could be accurately measured. There are two such possibilities. The first is based on the property of nitrosobenzene derivatives to form monomers after sublimation. Freshly sublimed crystals of monomer can be easily visualized because of their blue-greenish color. By removing the crystals from cold finger of the sublimer, the dimerization can be observed as a disappearance of the color. If the reaction is not too fast (or too slow) the crystals of monomers could be transferred to the infrared spectrometer, and the reaction followed by measuring the intensity of signals associated to N=O, or ON=NO stretching vibrations.

The second possibility for studying the solid-state dimerization is based on the producing monomers from starting crystal dimers by photodissociation. However, while the arrangements of the monomer molecules in the crystal obtained by sublimation and photolysis, respectively, can be different, the two approaches of studying the solid-state dimerization could deal with different reaction mechanisms. We will demonstrate, that combination of these approaches together with the previously represented knowledge about the reactivity of nitrosoaromates as nucleophiles, can give us a deeper inside in the reaction mechanisms of dimerizations of the nitroso compounds.

For studying chemical reactions and reactivity in solid state, two important evidences must be taken into account. First is so called topochemical condition for the activation of chemical process, i. e. the proper arrangement of reactants' molecules that lead to the formation of the corresponding transition state. Later we will demonstrate how such a condition influence on the formation of nitroso dimers. Secondly, the chemical processes in solid state include two types of transformations, chemical reaction, defined as a bond breaking/formation and phase changes of the crystal lattice [208]. To find out how a solid-state chemical reaction is coupled with phase conversion, it is appropriate to measure the chemical process and corresponding phase changes by different and independent methods. As we have demonstrated, the IR spectra of nitroso monomers and dimers are significantly different, especially in the appearance of new ON=NO stretching vibration that appears with the very sharp peak (at 1,260 cm^{-1}) in the spectrum. Following the change in intensity of this absorbance, it is possible to measure accurately the rate of chemical reaction. In practice, one obtains an exponential kinetic curve. On the other hand, the kinetics of the phase transformation can be unequivocally measured by time resolved X-ray powder diffraction (TR-XPRD) [209]. If the crystal structures of reactants, products, and perhaps intermediate metastable polymorphs are known, it could be possible to resolve the reaction mechanism, including the structure of the possible transition state. Most of the kinetic curves (the intensity of proper XPRD maximum plotted against time) of the solid phase transformations afford sigmoid shape, and were systematically analyzed by the traditional methods of M. Avrami and B.V. Erofeev [210–214]. Accordingly, by measuring independently these two processes by the time resolved IR spectroscopy, and by TR-XPRD, we have an experimental arrangement for resolving not only the reaction mechanism, but also the relation between two basic processes in the solid state [215].

Recent systematic investigations of dimerizations of *p*-substituted nitrosobenzene [215] demonstrated the role of the crystal packing on the rate and kinetics of dimerizations. Monomers of *p*-bromonitrosobenzene can be obtained *in situ* either by sublimation or by photolysis under the cryogenic conditions (under 170 K) [216].

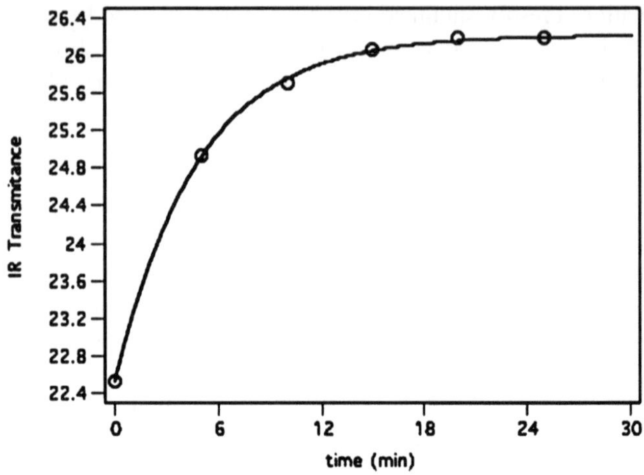

Fig. 3.5 Kinetics of the solid-state dimerization of *p*-bromonitrosobenzene recorded by measuring the ON=NO stretching signal at 1,260 cm^{-1} (Reproduced by permission from the Ref. [215])

If the synthetically prepared *p*-bromoazodioxide is sublimed, the blue-green crystals of the *p*-bromonitrosobenzene monomer appear on the cold finger of the sublimer. From the X-ray structural analysis of these crystals it follows that monomer molecules, which are packed with the stacked structure, are mostly arranged in such a way that nitroso groups of neighboring molecules are properly oriented to each other. Such an arrangement satisfies the *topochemical condition* for the reaction in which nitroso groups must become close.

At room temperature the color of the photolytically obtained monomer crystals disappears within 40 min because of the re-formation of azoxy dimers. If the process is measured by time resolved IR spectroscopy following the intensity of the new signal at 1,260 cm^{-1}, assigned to the ON=NO asymmetric stretching vibration, the reaction affords first-order kinetics with the rate constant $k = 3.48 \times 10^{-3}$ s^{-1} (25 °C), which corresponds to $t_{1/2} = 3.32$ min (Fig. 3.5). By application of the Arrhenius plot from the measurements at different temperatures, the activation energy of 145 kJ mol^{-1} has been calculated. Interestingly, it is higher than it has been found for similar reactions in solution. However, it must be mentioned that the described kinetics is not the representation of the "pure" chemical reaction but rather the combination of processes in which chemical reaction is one of the components. Kinetic measurements of the solid-state reactions suffer from the low accuracy. Dissipation of the rate constants could be as high as 30 %. Fortunately, recently we have found that the values of the kinetic isotope effect (k_{14N}/k_{15N}) have much lower dissipation of measured values. It could be speculated that the large deviation of rate constants origins from the influence of the phase change on the chemical reaction in the solid-state.

However, if the kinetic measurement is performed by the time resolved XPRD, shown on the next figure (Fig. 3.6), two different kinetic behaviors have been observed, depending on the X-ray reflection used.

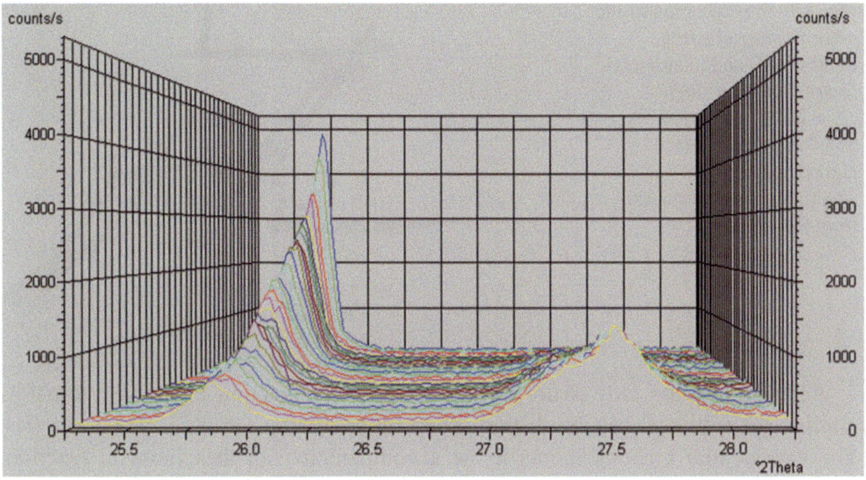

Fig. 3.6 Kinetics of the solid-state dimerization of *p*-bromonitrosobenzene recorded by the time-resolved XPRD, measuring the reflections at Θ = 27.5°, and Θ = 25.8° (Reproduced by permission from Ref. [215])

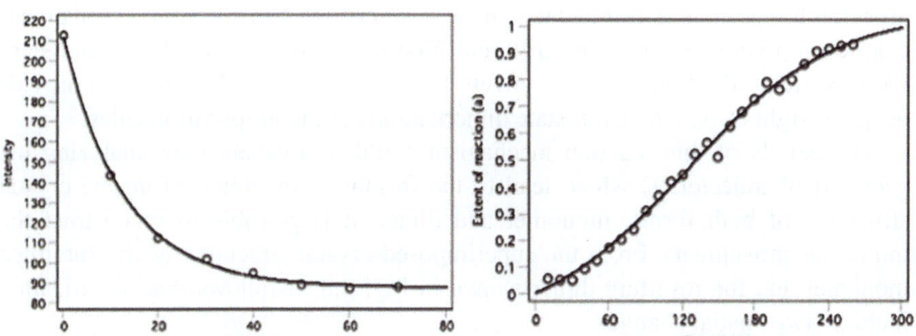

Fig. 3.7 Kinetic curves of the solid-state dimerization of *p*-bromonitrosobenzene recorded by the time-resolved XPRD, measuring the reflections at Θ = 27.5°, and Θ = 25.8° (Reproduced by permission from Ref. [215])

The reflection at the lower angle (Θ = 27.5°) diminished in its intensity following the exponential reaction path with the rate constant $k = 1.31\ s^{-1}$ (15 °C) that in principle corresponds to the rate constant obtained by the IR measurements (Fig. 3.7). Interestingly, this reflection corresponds to the crystal plane identified as [2.0.0.], which intersects the reactive nitroso groups of the neighboring molecules. The second reflection at Θ = 25.8° changes its intensity differently, after 40 min (when the reaction followed by IR is already completed) it starts to diminish relatively slowly affording the sigmoid-like kinetics (Fig. 3.7). Evidently, the second process is a phase transformation triggered by the dimerization reaction.

Fig. 3.8 Superposition of the *p*-bromonitrosobenzene dimers (*red*) and monomers (*green*) after the single-crystal-to-single-crystal transformation induced by photodissociation (Reproduced by permission from Ref. [218])

Accordingly, the two basic processes, chemical reaction and phase transformation, respectively, can be distinguished by analyzing the type of the kinetics. The exponential kinetics seems to be accompanied with the chemical reaction, and sigmoid-like kinetics is assigned to the phase change. In the case of *p*-bromonitrosobenzene the chemical reaction appears as a process within the first 40 min, and the phase transformation, governed by the growth of grains of new phase, takes another 300 min. Evidently, after 40 min, a metastable phase has been formed. Recent computer simulations of the crystal growth provided an additional aspect. It has been demonstrated that in the same crystal, two different crystal planes can develop simultaneously by different processes, exponential growth and step-wise by sigmiodal kinetics, respectively [217]. Perhaps this findings can provide deeper insight also in the solid-state dimerizations of the nitroso molecules.

The details of this reaction mechanism could be explained by analyzing the geometrical trajectories, which lead to the formation of dimer. From the crystal structures of both forms, monomer and dimer, it is possible to reconstruct the molecular movements. From the superimposed crystal structures of the sublimed monomer and the resulting dimer shown in Fig. 3.8, it follows that dimerization includes two spatial changes.

First movement is the twisting of the nitroso group plane out of the benzene ring plane, and the second includes the inclination of every second molecule of the dimer and the formation of a new packing plane. Tentatively, the first movement could be associated with the dominant chemical reaction measured as a 40 min process with exponential kinetics, and the second spatial shift corresponds to the slower 300 min phase transformation.

Freshly sublimed *p*-iodonitrosobenzene is also a colored monomer, but behaves differently. It remained in a form of the stable monomer molecule in the crystal even at room temperature [219], because the molecules in the crystal lattice are arranged in plane with the NO groups close to the iodine atoms [177, 220], and there are no neighboring pairs of nitroso groups, which can react with each other. This difference in packing of *p*-bromo- and *p*-iodonitrosobenzene molecules is shown on the Scheme 3.73.

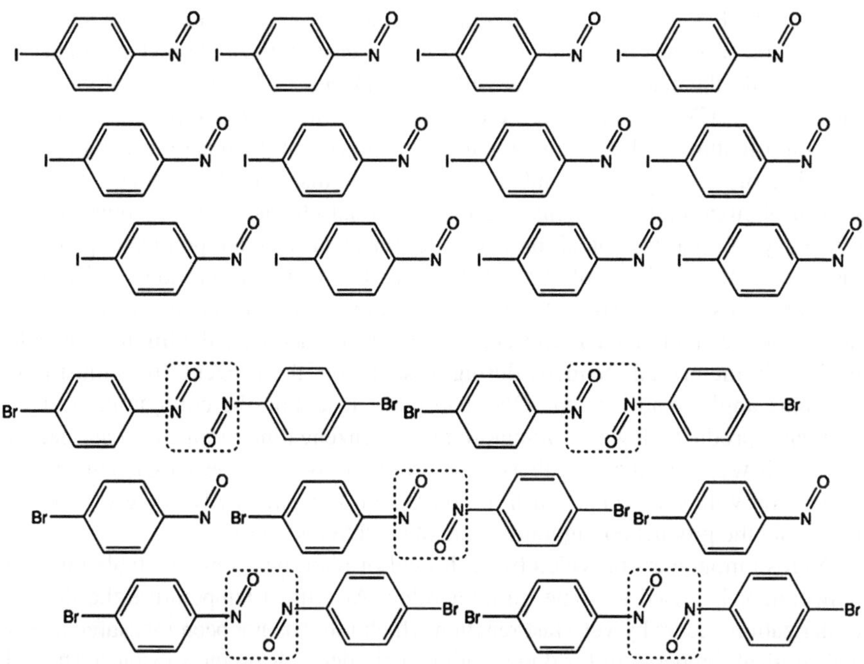

Scheme 3.73

Dimerization of the freshly sublimed *p*-methylnitrosobenzene is so fast (completed within 13 min) that only the sigmoid curve associated to phase transformation can be observed.

There is a long-standing discussion about the solid-state conditions that are necessary to trigger the reaction or phase transition. Scientists agree that the crystal lattice deformations serve as the points in the crystal where the reaction could start. If this is true, than the best positions on the crystal for initiating the process is on its surface. Since nitrosobenzene monomers readily sublime, it could be expected that sublimation of molecules from the surface forms deformations which can trigger the chemical process. Recently, it has been found that the previously discussed solid-state dimerization of *p*-bromonitrosobenzene is actually induced by sublimation [221].

Quite different behavior is observed if the monomer molecules are obtained by photolysis under cryogenic conditions [216]. The ultraviolet radiation in solution and at room temperature causes decomposition of most of the azodioxides by extrusion of NO [222–224]. Some dinitroso compounds undergo photorearrangements, as for instance *o*-dinitrosobenzene, which converts to benzofuroxan [225]. However, in solution at low temperature (−60 °C) it seems that some aromatic nitroso compounds undergo dissociation to monomers [175]. This kind of photodissociation is even more pronounced in the solid state at temperatures as low as 170 K, where a series of aliphatic and aromatic nitroso dimers dissociate to the corresponding monomers. The reaction can be visualised, because the obtained monomer is bluish while the starting dimer is yellow or colorless. This property of changing the color by UV irradiation represents a sort of the *photochromism*. Interesting structural feature of this nitroso photochromism is that the reaction includes only breaking of one covalent bond between two atoms (the N=N bond). In other known photochromic reactions, the constitutional change is more complex. The efficiency of such a photochromism in aromatic nitroso compounds depends on the nature and position of the substituents. For example, photodissociation of *p*-bromonitrosobenzene dimer is effective, but *m*-chloronitrosobenzene dimer does not dissociate at all. Such a difference in reactivity is associated with the molecular packing in the crystal. Namely, during dissociation, the benzene rings must twist for some angle relative to the ON=NO plane to obtain the configuration of the monomer product. However, *m*-chloronitrosobenzene dimer molecules are packed in such a way that the neighboring chlorine atoms form the Cl-Cl noncovalent interaction with each other. Such an interaction prevents the twisting of benzene rings, and the product configuration cannot be obtained [226].

Nitroso monomers provided by such kind of photolysis remain stable until the temperature is raised to some limiting value. At critical temperature the thermal dimerization occurs by very fast reaction which rate has not been measured as yet. This critical temperature for dimerization again depends on the substance studied. While *p*-bromonitrosobenzene redimerizes already at 170 K, some other derivatives return to the dimeric forms even close to 300 K [226]. It is of applicative interest to continue investigations in looking for the system that will work at room temperature.

From the mechanistic point of view, the photodissociation and subsequent thermal dimerization could provide new insight to the problem of monomer-dimer equilibrium. If the dimer of *p*-bromonitrosobenzene in single crystal is photolysed at 100 K inside the X-ray diffractometer, the single-crystal-to-single-crystal transformation of the dimer to corresponding monomer has been observed [218].

As it could be seen in Fig. 3.8, in which the photoproduct monomers are superimposed to the starting dimers, the molecules of monomeric units (green) undergo drastic reorientation in comparison with the starting dimers (red). This molecular rearrangement includes also tilting of the benzene ring. The question that appears is why, in spite to the large molecular reorientation, the reverse dimerization occurs so quickly, and already at the temperature as low as 170 K. The solution could be found by looking in the details of crystal packing.

In the metastable phase that appears immediately after photolysis (represented in Fig. 3.9), the monomer molecules remain in the vicinity of each other with the extraordinary close contact between nitrogen atoms of the neighboring nitroso groups. The N, N distance of 2.30 Å is for the 23.3 % shorter than the sum of the van der Waals radii (3.0 Å) of two nitrogens (Scheme 3.74). The shortest N, N close contact between molecules other than nitroso compounds found in literature is 2.62 Å. The close contact between the nitrogen atom of one molecule and the oxygen of the nitroso group of neighboring molecule, 2.35 Å is also very short. From the quantum chemical calculations by using density functional theory (B3LYP with

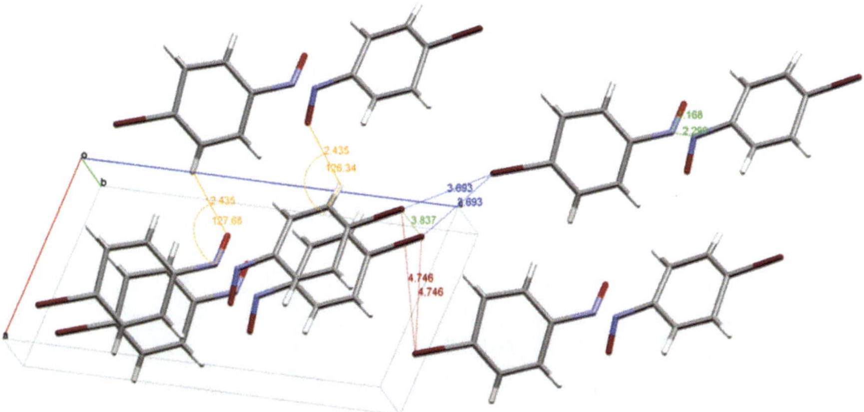

Fig. 3.9 Molecular packing of the *p*-bromonitrosobenzene molecules after photodissociation. The 2.30 Å close contact between two N-atoms is labeled in *green color* (Reproduced by permission from Ref. [218])

Scheme 3.74

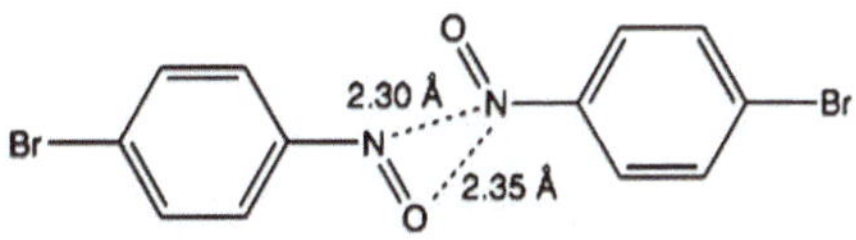

the 6-311+G(d,p) basis set), it was found that such closely contacted molecular pair of nitroso monomers is not persistent in the gas phase [218], and that relatively high charge density remains between the neighboring nitrogen atoms.

The appearance of such a short contact could give explanation why the dimerization under these conditions in the crystal is so fast even at 170 K. However, the question arises, what would be the mechanism of dimerization governed by such a dense intermolecular arrangement? It is difficult to believe that the mechanism is the same as in the case of dimerization of monomers obtained after sublimation.

In discussed examples, the efficiency of dimerization depends mostly on the stereoelectronic structure of monomer molecule and on the conditions of the immediate environment. Solid crystal phase promote formation of dimers perhaps because of fixed positions of molecules as well as because of properly oriented nitroso groups with close contacts. It would be interesting to investigate whether the molecular self-organization in only two dimensions could also promote dimerization. Nitrosobenzene molecules with the long aliphatic chain ending with SH or SCN groups bound to the *para* position can form a self-assembled monolayer (SAM) [227] on the regular Au(111) surface, as it is shown in Fig. 3.10.

Molecules in the SAM are organized in hexagonal 2D structures with NO groups on the surface. Such a monolayer surface serves as a matrix for the arrangement of the second layer by forming the ON=NO bonds. Such a self-assembled bilayer (SAB) (Fig. 3.11) consisting of two layers of self-organized

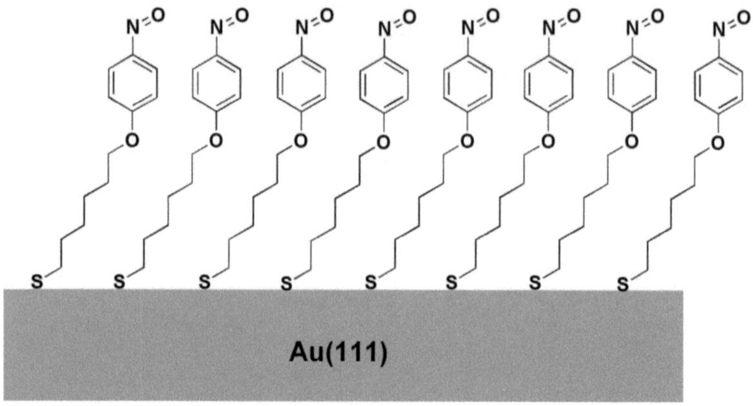

Fig. 3.10 Self-assembly of the nitrosoaromatic molecules on the regular Au surface

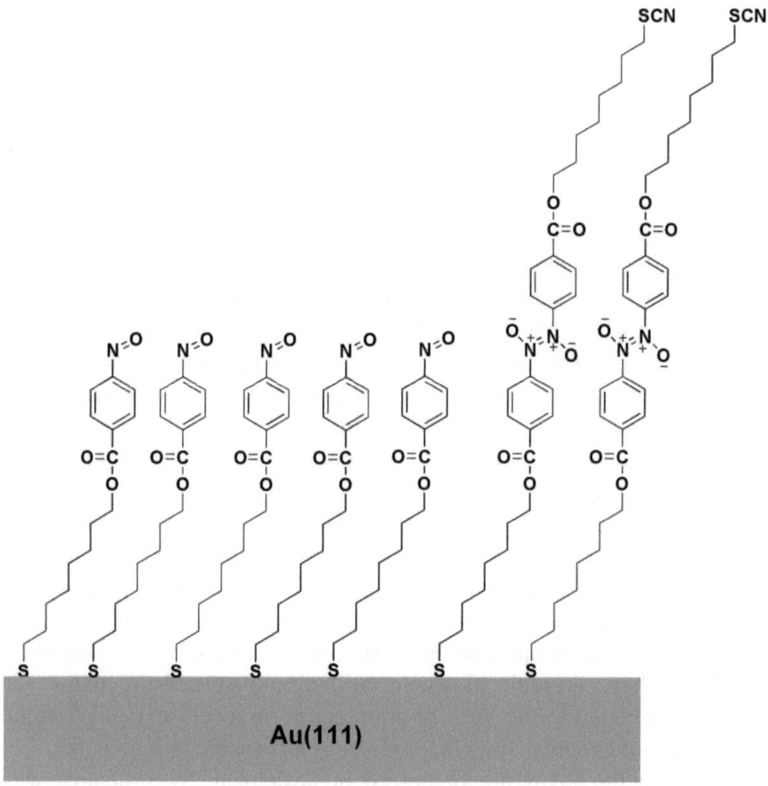

Fig. 3.11 Self-assembled bilayer (SAB) obtained by formation of the azodioxide bonds

Fig. 3.12 STM image of the self-assembled bilayer (SAB) obtained by formation of the azodioxide bonds (Reproduced by permission from Ref. [227])

Scheme 3.75

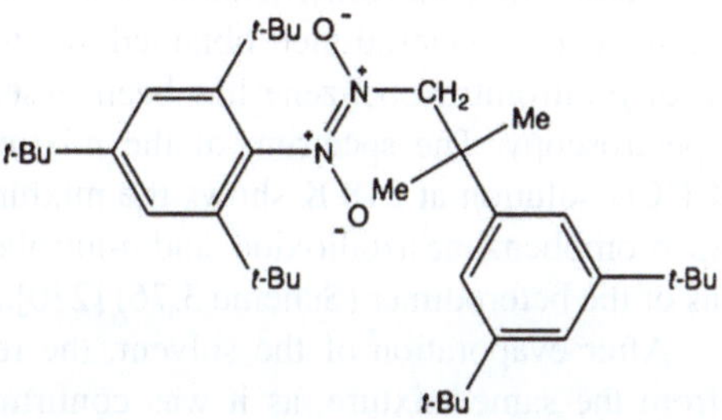

monolayers interconnected by azodioxide bonds shows that 2D spatial organization of molecules strongly promotes the dimerization. The regular surface of SAB can be recognized on the STM micrograph (Fig. 3.12) [227].

3.3.4 Cross Dimerizations

In the mixture of two different nitroso compounds, say A-NO and B-NO, formation of azodioxide could include four combinations of nitroso precursor: A-(O)N=N(O)-A, B-(O)N=N(O)-B, A-(O)N=N(O)-B, or B-(O)N=N(O)-A. While first two are symmetrical, let's call them *homodimers*, the last two are cross-linked, let's call them *heterodimers*. In first observed heterodimers known from the literature, the A-side is aromatic, with the voluminous substituents in *ortho* position, and the B-side is aliphatic, and not sterically hindered [228]. Their appearance is explained by inability of the aromatic nitroso component to form homodimers because the large *ortho* substituents (*tert*-butyl groups in Scheme 3.75) prevent the molecules to approach to each other.

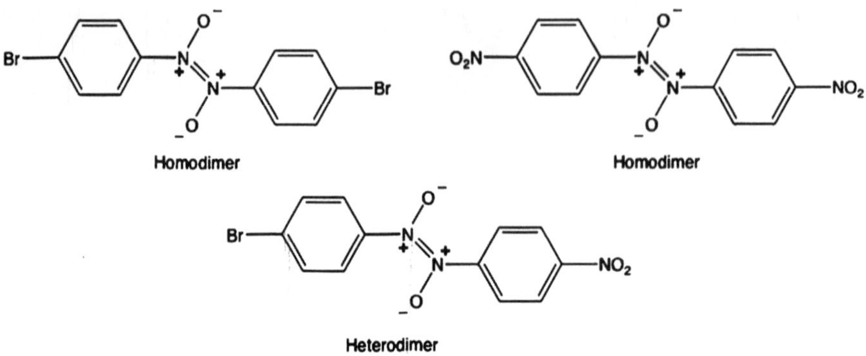

Scheme 3.76

A series of such heterodimers with the sterical hindering of *ortho* groups were prepared [229].

From mechanistic point of view it is much more interesting to examine selectivity in cross-dimerizations that depends solely on the electronic effects. First such a heterodimer obtained by the reaction of *p*-bromonitrosobenzene with *p*-nitronitrosobenzene has been observed in solution by (^{1}H) COSY NMR spectroscopy. The spectrum of the mixture of these two precursors recorded in $CDCl_3$ solution at 218 K shows the mixture that consists of both the homodimers (*p*-bromobenzene azodioxide, and *p*-nitrobenzene azodioxide, respectively) as well as of the heterodimer (Scheme 3.76) [230].

After evaporation of the solvent, the remained polycrystalline solid consisted from the same mixture, as it was confirmed by CP MAS NMR spectroscopy as well as by X-ray structure determination. Both the homodimers, together with the heterodimer, form cocrystal with the molecular arrangement similar to that of the crystal of *p*-bromonitrosobenzene dimer. If the cocrystals are prepared from the mixture in which *p*-bromonitrosobenzene was labeled with ^{15}N, three signals assigned to the ON=NO vibration were observed, $^{14}N^{14}N$ at 1,264 cm^{-1} (*p*-nitro homodimer), $^{15}N^{15}N$ at 1,234 cm^{-1} (*p*-bromo homodimer), and $^{14}N^{15}N$ signal at 1,246 cm^{-1}, associated to the heterodimer [230].

A series of *p*- and *m*- substituted nitrosobenzenes were studied regarding the selectivity in the formation of heterodimers [231]. If the parent nitrosobenzene was used as one side of the heterodimer, variation of the second partner can give us some measure for the selectivity in formation of heterodimers (Scheme 3.77).

It was found that *p*-nitronitrosobenzene forms the most stable heterodimers, and that this derivative can combine with all the investigated partner molecules, even with *p*-methoxynitrosobenzene, for which it is known that in principle does not form homodimers at all. Remember, that *p*-nitronitrosobenzene is also one of the most reactive targets for nucleophilic attack. This evidence, together with the presented knowledge about reactivities for the formation of heterodimers, could be an indication that the mechanism of dimerization includes nucleophilic addition of one nitroso molecule to another with nucleophilic N-atom.

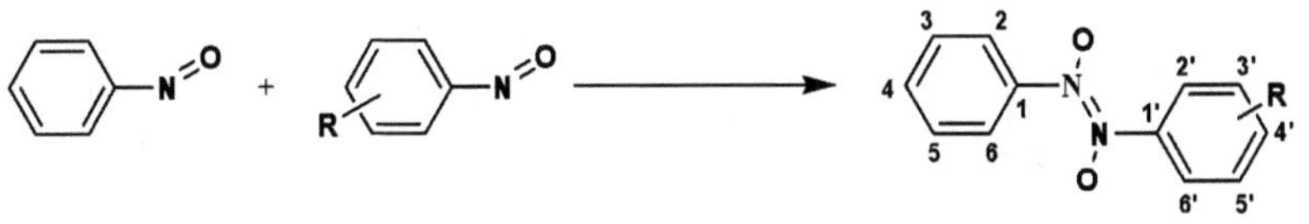

1, R = p-H
2, R = p-CH_3
3, R = p-F
4, R = p-Cl
5, R = p-Br
6, R = p-I
7, R = m-Cl
8, R = p-OCH_3
9, R = p-NO_2
10, R = p-$COOCH_3$
11, ^{15}N-nitrosobenzene

Scheme 3.77

Table 3.20 Formation of heterodimers in solution and in solid-state

Combination	Solution	Solid-state
1 + 2		X
1 + 3		X
1 + 4		X
1 + 5	X	X
1 + 6	X	X
1 + 7	X	X
1 + 8	–	–
1 + 9	X	X
1 + 10	–	–

In solid-state, the appearance of heterodimers is different than in solution (Table 3.20). Most of the *E*- isomeric heterodimers that are not persistent in solution exist in solid state. This confirms our previous assertion that molecular self-organization, either in two or in three dimensions promote stabilization of dimers.

3.3.5 Dimerizations of Heteroaromatic Derivatives

The model compound for studying heteroaromatic derivatives is pyridine. In contrast to the substituted nitrosobenzenes, which are monomeric in solution under the standard conditions, dissolved 2-nitrosopyridine appears as the *Z*-dimer already at room temperature [168]. High stability of the dimer has been explained by interactions of the pyridine rings in the *cis* form. However, the additional explanation could be also based on the instability of the monomer that has a partial nature of the nitrene cation like form, as it could be seen from the resonance structures (Scheme 3.78).

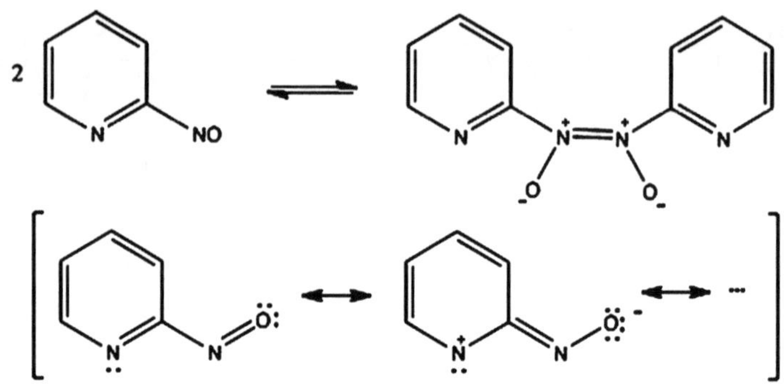

Scheme 3.78

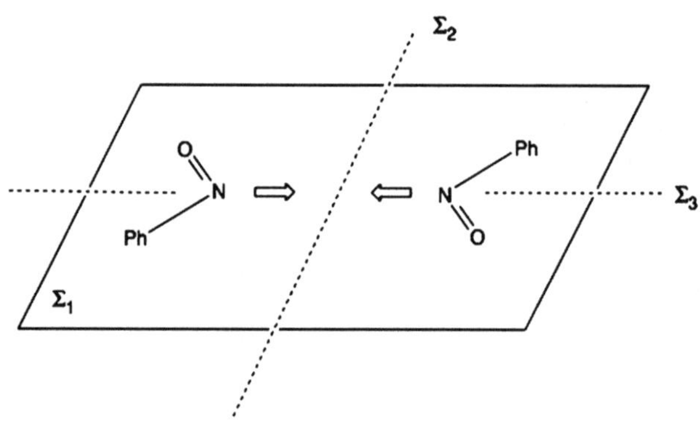

Scheme 3.79

Since nitrosofuran as well nitrofuran are unstable molecules their experimental study is rather difficult [223]. The molecular structure of 5-nitrosofuran has been studied by quantum chemical calculations [232]. It has been found that the molecule prefers the dimeric form with the *Z*-isomer slightly (nearly for 2 kJ mol^{-1}) more stable than the *E*-isomer. The homoaromatic analog, 5-nitroso-2,3-cyclopentadiene has not been isolated, but from the calculations it follows that its structure is monomeric [233].

3.3.6 Calculations of the Reaction Path

Two possible reaction paths by which the monomer molecules approach to each other to form dimers are possible. The simplest one is the *least-motion path* where both the monomer molecules draw closer in a plane, as it is described in the Scheme 3.79.

In this path, firstly studied by Roald Hoffmann et al. [234], it is possible to construct a sort of level correlation diagram, normally applied for studying concerted reactions such as 2p + 2p addition of the two singlet carbene molecules.

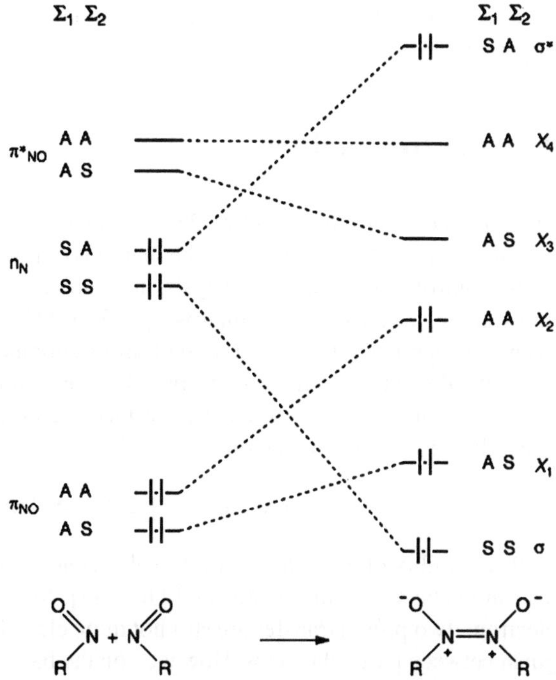

Fig. 3.13 MO diagram for dimerization of nitroso molecule models (Adopted from Ref. [234])

On the reactant, the nitroso monomer side, the relevant orbitals are two n-lone pairs as well as the π and π^* on the NO group. On the dimer side there are N-N σ-orbital, and four π orbitals. Approximating this least-motion approach with the D_{2h} symmetry, these orbitals can be arranged regarding their symmetry to the planes Σ_1, and Σ_2, respectively, (Scheme 3.79). On the correlation diagram (Fig. 3.13) the label S represents the symmetric, and the symbol A the antisymmetric property in relation to the Σ_1 and Σ_2 planes.

From the correlation diagram it is clear, that transformation of the reactants to the products should include excited states. The n_N orbital of the SA symmetry of the reacting monomers appears as σ^* SA orbital in the dimer. Accordingly, the least-motion path is symmetrically forbidden, and there is only the possibility of a photoreaction. However, the detailed study of the correlation state diagrams has shown, that even in the case of photoreaction there is no such a path, which satisfies the conservation of the D_{2h} type of symmetry [235].

The dimerization possibly occurs by using a more relaxing non-least-motion way. W. Lüttke et al. [236] have run *ab initio* calculations for the model system of dimerization of the parent structure HNO. As the previous studies suggested, the path includes the stationary point, the transition state, with the molecules approaching "side-to side":

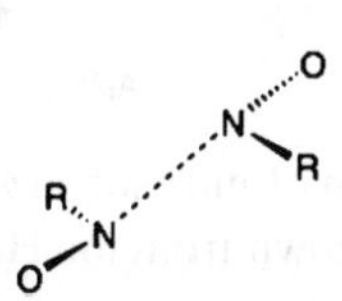

The N-N distance in the calculated transition state is 1.852 Å for HNO dimerization.

3.4 Electrochemistry

Most of recent investigations [39, 237] of the electrochemical behavior of aromatic nitroso compounds were based on the methods of polarography, cyclic voltammetry [238], controlled potential electrolysis, in combination with EPR spectroscopy [239] and scanning electrochemical microscopy [240, 241]. Electrochemical synthesis has even been applied in the preparation of nitroso compounds [242–245]. In principle, two typical reaction paths were proposed, dependently on the solvent in which the reaction occurs. In protic solvents, the mechanism depends on pH, but could be generally described by equation:

$$\mathrm{ArNO} + 2\mathrm{e}^- + 2\mathrm{H}^+ \rightleftharpoons \mathrm{ArNHOH}$$

Reversibility of the redox pair nitrosobenzene-phenylhydroxylamine is observed also in cyclic voltammetry studies [246, 247]. Mechanistic details of such a two electron - two proton transfer are still not quite clear, because it is difficult to distinguish between particular steps. However, on the basis of a good agreement between the simulated and experimental cyclic voltammograms, the mechanism that involves four stages could be proposed (see following equations) [239]. In the first step, the electron capturing by nitroso molecule results with the formation of radical anion. Proton donors transform the anion radical to the corresponding hydronitroxide. Such protonation occurs regularly during the electrochemical reductions of nitroso compounds [248]. All the electrochemical steps are fast and reversible with the easier transfer of the second electron.

$$\mathrm{ArNO} + \mathrm{e}^- \rightleftharpoons \mathrm{ArNO}^{\bullet -}$$

$$\mathrm{ArNO}^{\bullet -} + \mathrm{H}^+ \rightleftharpoons \mathrm{ArNOH}^{\bullet}$$

$$\mathrm{ArNOH}^{\bullet} + \mathrm{e}^- \rightleftharpoons \mathrm{ArNOH}^-$$

$$\mathrm{ArNOH}^- + \mathrm{H}^+ \rightleftharpoons \mathrm{ArNHOH}$$

It has been found that the reaction system is even more complicated and strongly dependent on pH [249]. Further protonation of the resulting hydroxylamine derivative yields aniline.

$$\mathrm{ArNHOH} + \mathrm{H}^+ \longrightarrow \mathrm{ArNH_2OH}^+$$

$$\mathrm{ArNH_2OH}^+ \xrightarrow{+\,2\mathrm{e}^-,\ +\,\mathrm{H}^+} \mathrm{ArNH_2} + \mathrm{H_2O}$$

The sensitivity of the reduction of nitrosobenzene to the substituents in the *para* position is negligible, as it was shown from the Hammett correlations [39].

Scheme 3.80

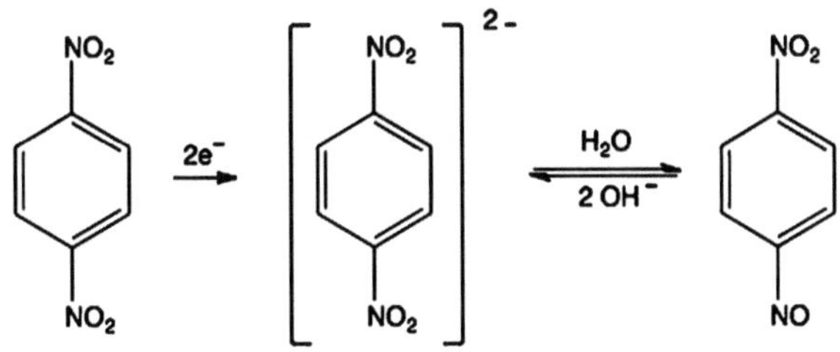

In aprotic media, the dominant process is one electron transfer with the formation of a nitroso radical anion that could be combined with the components of the solvent to form ion pairs. Additionally, there is also the competitive condensation reaction leading to the azoxide product. The overall reaction path includes two irreversible parallel reactions: dismutation to the product pair nitroso compound and hydroxylamine, and dimerization to azoxybenzene:

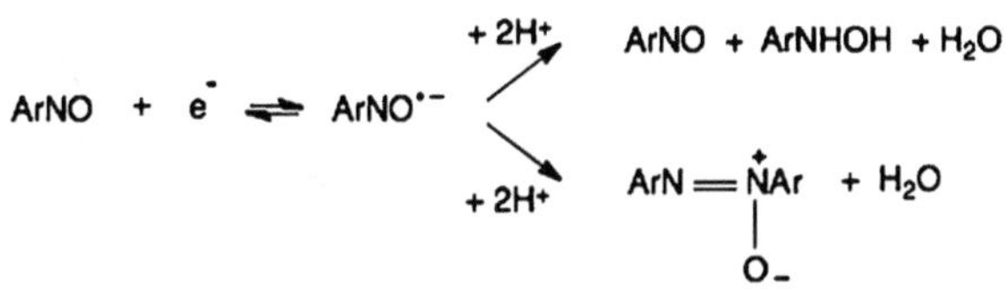

Appearance of the nitrosobenzene radical anion has been confirmed by EPR spectroscopy [250] in the reaction of nitrosobenzene with glutathione. Nitrosobenzene radical anion can under aerobic conditions react with oxygen molecule to afford oxygen radical anion.

$$ArNO^{\bullet-} + O_2 \rightarrow ArNO + O_2^{\bullet-}$$

This reaction may be *inter alia* responsible for the toxicity of nitrosobenzene, because the oxygen radical anion can damage cell membranes.

Recent studies of the electron transfer rate constants in reductions of the dihydropyridine substituted nitrosobenzenes by scanning electrochemical microscopy (SECM) have informed about strong dependence of the heterogenous reductions on the pH and the electrode material [240, 241].

In recent studies, the redox systems nitrobenzene-nitrosobenzene have been investigated extensively by combined methods of FT-IR spectroscopy and cyclic voltammetry [251]. Reduction of 1,4-dinitrobenzene by the two-step process that includes the transfer of two electrons, leads to the formation of the unstable dianion, which after protonation forms 1-nitro-4-nitrosobenzene (Scheme 3.80). The nitroso product has been identified by *in situ* FT-IR spectrum where the absorption band at 1,587 cm^{-1} has been assigned to the nitroso N=O stretching vibration.

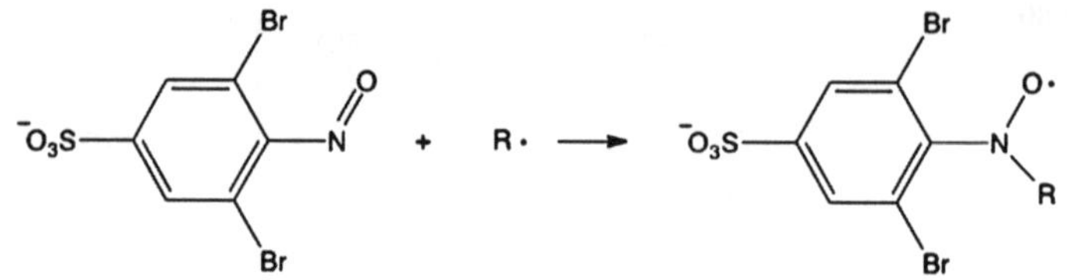

Scheme 3.81

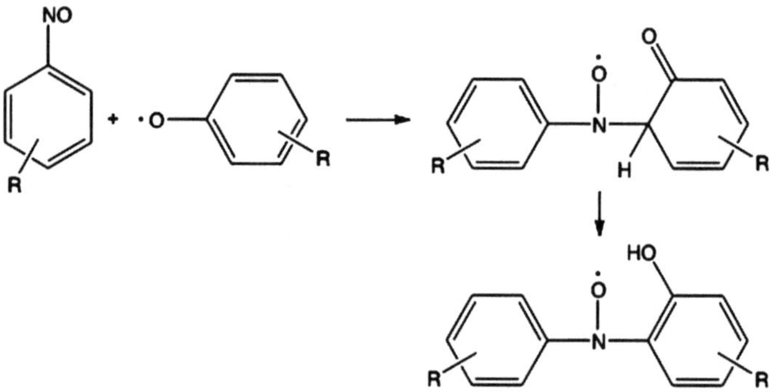

Scheme 3.82

3.4.1 Aromatic Nitroso Compounds as Spin Traps

Spin trapping is analytical method for indirect detection of radicals, either in EPR spectroscopy or in electrochemistry. Reagent used for spin trapping reacts with the unstable radical under the investigation, and forms the radical that is enough persistent to be measured. Numerous spin traps are known and summarized in literature [252]. Besides nitrones, nitroso compounds serve as standard spin traps. The most popular nitroso spin traps for the EPR spectroscopy are 2-methyl-2-nitrosopropane (MNP), and the aromatic nitroso derivatives nitrosobenzene (NB), 2,3,5,6-tetramethylnitrosobenzene (nitrosodurene, ND) [253], and 3,5-dibromo-4-nitrosobenzene sulfonate (DBNBS) [254]. Basically, the radical intermediate in question adds to the N-atom of the nitroso group forming the relatively stable nitroxide radical (Scheme 3.81).

The addition of radicals with localized electron mostly yields only one spin trap, as it is described on the Scheme 3.81. However, the mechanism of addition of aromatic radicals with delocalized unpaired electron is more complex because the bond formation with the molecule that serves as a trapping reagent could be possible on different positions on the benzene ring. It has been calculated that the selection of the atom on the radical molecule that will form the chemical bond with the nitroso N-atom *inter alia* depends on the spin density [255, 256]. In the case of phenoxyl radical, the highest spin density is calculated to be in *o*- and *p*-positions [253]. If the phenoxyl radical is sterically hindered with substituents in *p*-positions, the nitroxide radical is formed with the attack on the *o*-carbon (Scheme 3.82).

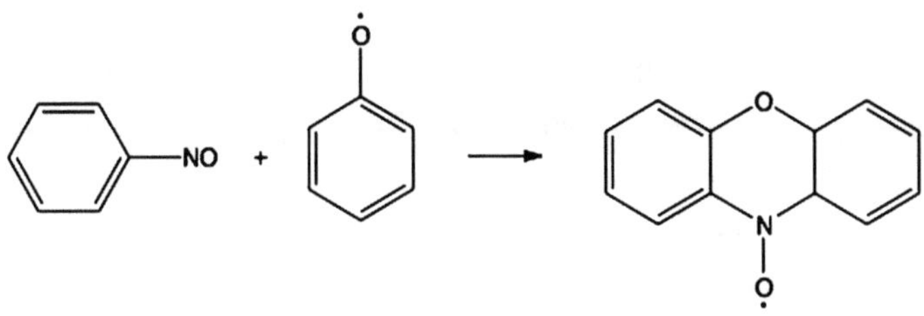

Scheme 3.83

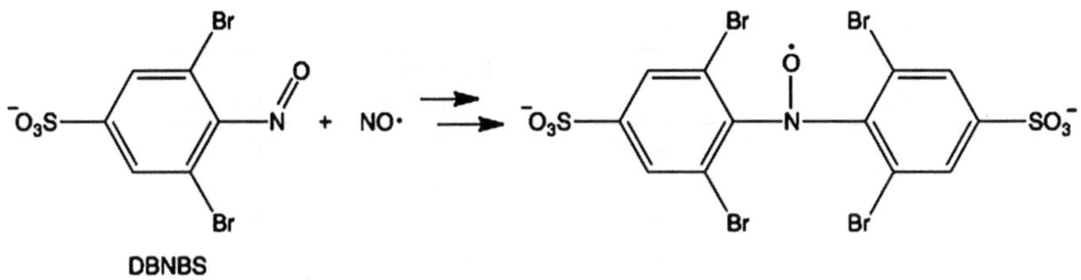

Scheme 3.84

Spin trapping of the parent nitrosobenzene with phenoxyl radical lead to cyclization and formation of phenoxazine-10-oxyl radical (Scheme 3.83) [253].

The DBNBS spin trap has its potential application for detection and analysis of nitric oxide, NO, the biologically interesting molecule. In organisms, the NO molecule is released by action of so called endothelium-derived relaxing factor (EDRF), which acts during the response of living cells on the cytotoxic stimulations [257]. Because of its short half-life, direct measuring of nitric oxide radical is complicated. The indirect detection by using DBNBS has been studied by Davies et al. [258]. Trapping of NO with 3,5-dibromo-4-nitrosobenzene sulfonate results with the formation of the stable radical dianion bis (2,6-dibromo-4-sulfophenyl nitroxyl), which can be detected by EPR spectroscopy (Scheme 3.84). It must be pointed out that the dianion radical is not a direct product of the NO trapping but the result of a more complicated mechanism.

In principle, the N-attack to the nitroso group appears as the first step before the formation of the dianion (Scheme 3.85).

The EPR spectral data for a series of spin adducts of DBNBS with radicals, which regularly appear in living cells are now available [259].

The use of DBNBS in analysis of radicals in biological samples such as human plasma could be limited because of its side reactions with the oxidants and formation of radicals with different and still unknown structures [259].

After the pioneering work by Bard [260], the use of aromatic nitroso compounds as spin trapping reagents in electrochemistry has been systematically investigated [261]. The main condition, which the spin trapping reagent must satisfy, is the potential domain where spin trap is neither oxidized nor reduced at the electrode

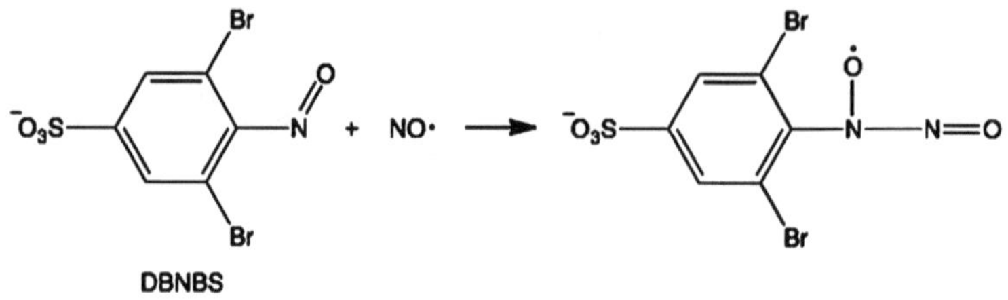

Scheme 3.85

Table 3.21 Dissociation of substituted nitrosobenzenes measured by cyclic voltammetry

Compound	Dissociation coefficient (%)	References
2,6-dichloronitrosobenzene	30.80	[264]
2,4,6-trichloronitrosobenzene	51.80	[264]
2-methylnitrosobenzene	99.15	[186]
4-methylnitrosobenzene	99.97	[175]
2,6-dimethylnitrosobenzene	22.40	[174]
2,4,6-trimethylnitrosobenzene	51.60	[264]
2,3,5,6-tetramethylnitrosobenzene	6.27	[264]
Pentamethylnitrosobenzene	8.13	[264]
Pentamethoxynitrosobenzene	21.00	[264]

surface. Such a domain is commonly called the potential window [262, 263]. The width of the potential window depends on the substituents on the benzene ring, and ranges from 2.3 V for pentamethylnitrosobenzene to 2.8 V in the case of the parent nitrosobenzene.

Complication in using nitrosobenzenes as electrochemical spin traps is also the monomer-dimer equilibrium. Consequently, the second important parameter that must be known in order to apply nitroso spin traps is the dissociation constant.

Of the two current peaks in the reduction of nitrosoaromatic compounds, the reduction potential at −0.61 V corresponds to the monomer, and the second maximum at −1.10 V is attributed to the reduction of the dimer. While the dimer radical anion has transient character, the monomer radical anion is the stable species. Difference in these current peaks has been used for measuring the dissociation coefficients for a series of aromatic nitroso compounds [264]. The results obtained at 298 K in the 0.001 M solutions are shown in the Table 3.21.

Use of the nitroso compounds as spin-trapping reagents in medical examinations must be taken with care because of their cytotoxicity. Among the frequently applied spin traps, the nitroso derivatives afford the highest toxicity [265]. For instance, the IC_{50} values (the concentration of the trapping substance that kills 50 % of cells) for mostly used nitroso spin traps are below 0.7 mM. For comparison, in the Table 3.22 is represented also the IC_{50} for the spin trap DMPO that does not belong to the nitroso derivatives.

Table 3.22 Toxicity of different spin trapping substances (Data from Ref. [265])

Compound	IC_{50} value	IC_{10} value
2-Methyl-2-nitrosopropane (MNP)	0.1	0.007
Nitrosobenzene (NB)	0.06	0.006
3,5-Dibromo-4-nitrosobenzenesulfonic acid (DBNBS)	0.64	0.07
5,5-Dimethylpyrroline-1-oxide (DMPO)	138.34	86.8

Scheme 3.86

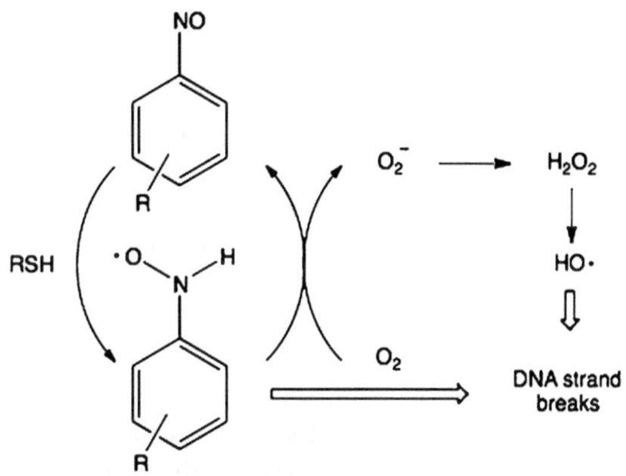

It is well known that the nitrosoaromatic spin traps in the living cell have substantial influence on the DNA strand break in the presence of thiols (for example the amino-acid cysteine). The mechanism is still not clear, but it is known that oxidation of O_2 to O_2^- as well as contribution of H_2O_2 and metal ions take a role in the process [266]. The tentative reaction path is shown in the Scheme 3.86.

Additional disadvantage in using nitrosoarene as spin traps is that they can not be used for detection of NO_2 because the formed adducts are unstable and cannot be unequivocally detected. In fact, the appeared adducts could easily react with another NO molecule in the reaction that is faster than trapping [267].

3.5 Photochemistry

Photoreactions of nitrosobenzene are known already more than a hundred of years ago. Bamberger (1902) reported first observations of the nitrosobenzene reactivity triggered by sunlight [268]. Since the nitroso compounds absorb in two spectral regions, in UV between 200 and 350 nm, and in visible between 700 and 800 nm, respectively, it could be expected that irradiation by both wavelengths may result with the chemical transformation. However, aliphatic and aromatic nitroso compounds afford different behavior in respect to the wavelength selected for the photochemical reaction [269]. Generally, aliphatic nitroso derivatives are more

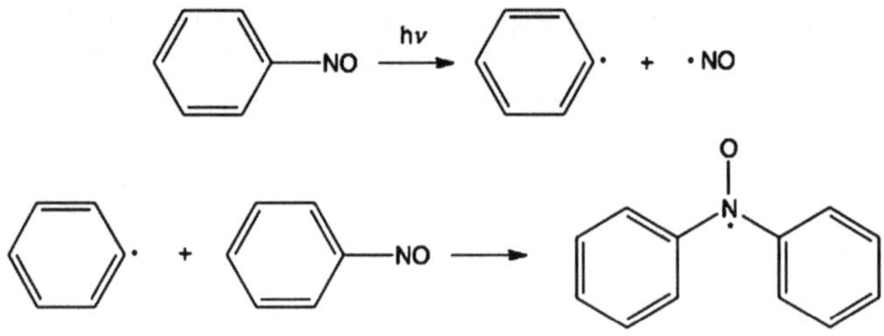

Scheme 3.87

reactive to the visible light than aromatic nitroso compounds. The difference could be explained by interactions of n-orbitals on the nitrogen and oxygen atoms with the rest of the molecule. Obviously, the interaction is only possible with the aromatic π-systems, which are absent in alkylnitroso compounds. Such explanation is supported by investigation of photoreactions of more or less sterically hindered nitrosobenzene derivatives, where the strong hindering reorients the NO group relative to the planar aromatic ring [228].

In early studies of the photochemistry of nitroso compounds, two main reaction types were recognized, formation of the aryl and NO radical pair, and photodissociation of dimerized form of the nitroso compounds in solution at lowered temperatures. While the first reaction is interesting for medicinal chemistry, the second could have application in the development of novel photochromic systems, as we have already discussed in Sect. 3.3.3.

Cleavage of the CN bond in aliphatic and aromatic nitroso compounds by UV-visible irradiation is already known for 50 years [270, 271]. It was found that the photolysis of nitrosobenzene in solution yields a complex mixture of products [272]. The proposed mechanism includes formation of diphenylnitroxide radical as a reactive intermediate. Appearance of nitroxide has been explained by reaction of phenyl radical with the nitrosobenzene molecule. Consequently, the first step in this process must be homolytic cleavage of the C-N bond (Scheme 3.87).

The presence of the diarylnitroxide molecules was confirmed by EPR spectroscopy and by studying the reactions in the benzene and deuterated benzene solutions [273].

The C-N bond in nitroso compounds is relatively week and can be easily broken by electronic excitation (Scheme 3.88). Its strength in nitrosobenzene has been found to be 215 kJ/mol [274, 275], and calculated as 228 kJ/mol [276] so that the wavelength for bond dissociation must be lower than 500 nm. The EPR spectroscopy and photosensitization experiments suggested that the dissociation to phenyl and NO radicals occur via an excited singlet state [277]. From studies of the UV-photodissociation of jet-cooled nitrosobenzene [278] follows that the required irradiation must be sufficient for excitation in the higher singlet states. The excess of the irradiated energy (that appears as a difference between the absorbed one and the required for dissociation) is redistributed between rotational and translational motions.

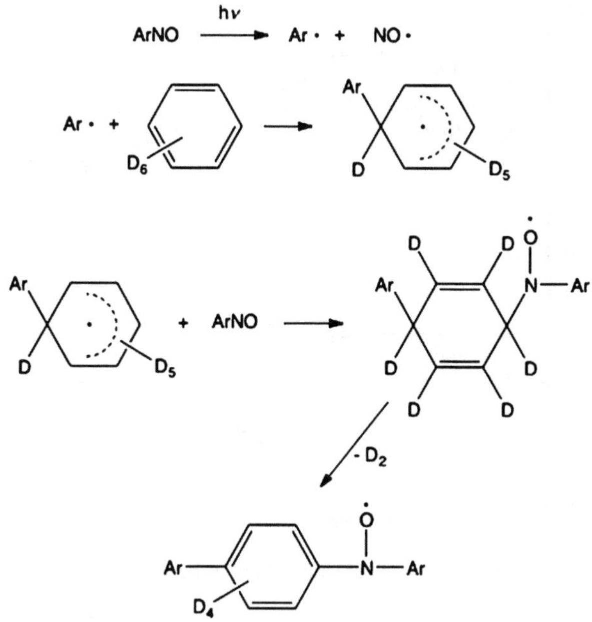

Scheme 3.88

Detailed IR and UV-VIS [154, 279] spectroscopic investigations of such nitrosobenzene fragmentations have been performed by photolysis in argon matrices. From both the spectroscopic methods, the phenyl radical and NO have been found to be the main products. Hacker et al. [280] have analyzed the EPR spectra of the argon matrix isolated nitrosobenzene at 12 K after the UV photolysis. From the comparison of the experimental and simulated spectra, they have concluded that the photodissociation yields three species: the isolated phenyl radical, and two triplet radical pairs Ph . . . NO. The two triplet radical pairs differ in the distance between Ph and NO radical, which can be 430 or 620 pm.

The overall mechanism of the nitrosobenzene photodissociation was proposed on the basis of the velocity-mapped ion imaging [281]. The initial absorption at 305 nm excites PhNO to the second singlet state S_2. Since the lifetime of S_2 is only 100 fs [154], it is too short for the dissociation to occur. Rather, the molecule decays to S_1 or even to the ground S_0 state by internal conversion. On the basis of the conservation of orbital symmetry rules, the dissociation is preferred from the ground, the S_0 state. It must be noted that other nitroso compounds could undergo also unimolecular photodissociation through excited triplet states [282–286].

The early systematic studies of the photoreduction of nitrosobenzene in methanol have demonstrated that after the excitation to the singlet state, the intersystem crossing convert the molecule to triplet state. These observations were confirmed by product analyses as well as by EPR spectroscopy [287]. Photoreduction in propan-2-ol follows the same photophysics [288].

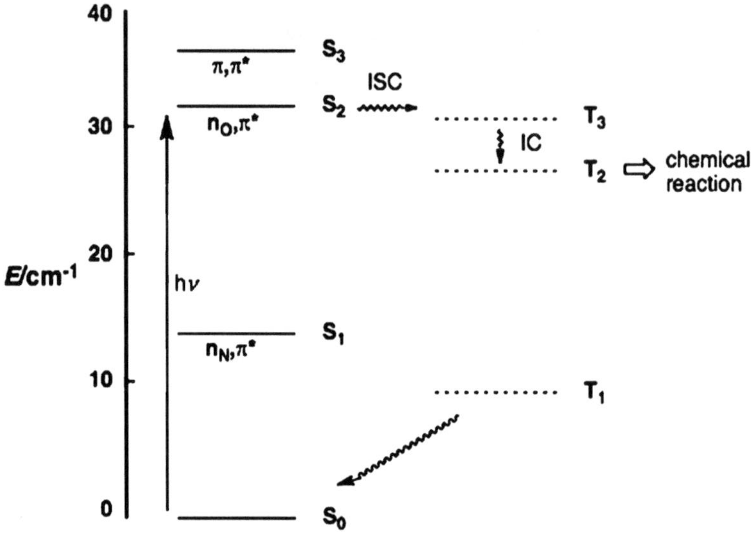

Fig. 3.14 Energy diagram for the photoreduction of nitrosobenzene

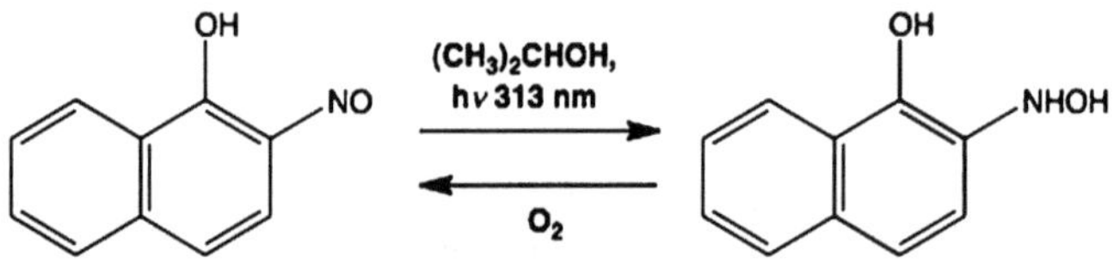

Scheme 3.89

In contrast to the aromatic ketones where $n\pi^*$ and $\pi\pi^*$ triplet states are close to each other, the energy difference of two triplets (T_1 and T_2) in aromatic nitroso compounds is much larger. The consequence of this separation is that different processes can be operative from these excited states, as it is shown on the energy level diagram (Fig. 3.14) [289].

While from the lowest triplet, T_1, the molecule returns by radiationless transition to the ground state, chemical reaction occurs from the higher T_2 state.

Main products of the photoreduction of aromatic nitroso compounds are corresponding hydroxylamines. In the presence of oxygen, the dark reaction after photolysis reconverts hydroxylamine to the starting nitroso derivative [288]. Obviously, hydroxylamine and nitroso derivative can also in the dark reaction condense to the azodioxide byproduct. An illustrative example of such a photoreduction is reaction of 2-nitroso-1-naphtol (Scheme 3.89) [289].

In his first paper about the nitrosobenzene photochemistry from 1902, Bamberger [268] reported that the longer exposure of nitrosobenzene to sunlight results with the complex mixture of the products. Azoxybenzene, 2-hydroxyazoxybenzene, and nitrobenzene were isolated as the main components of the mixture. For the

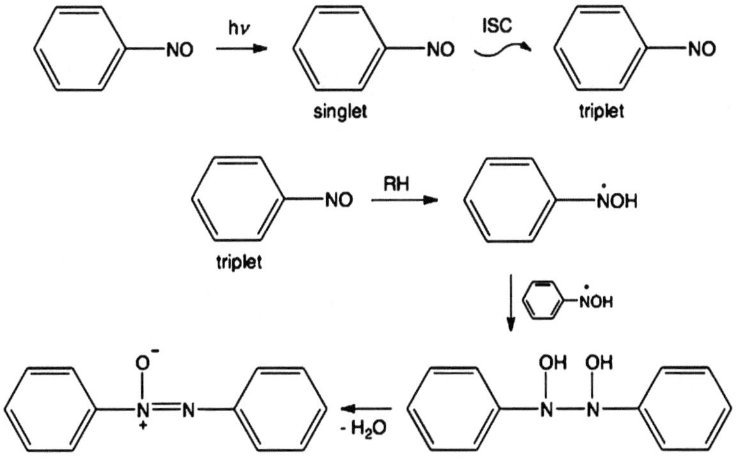

Scheme 3.90

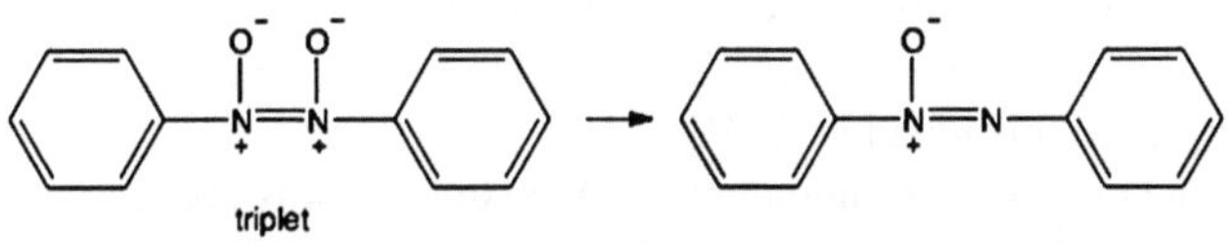

Scheme 3.91

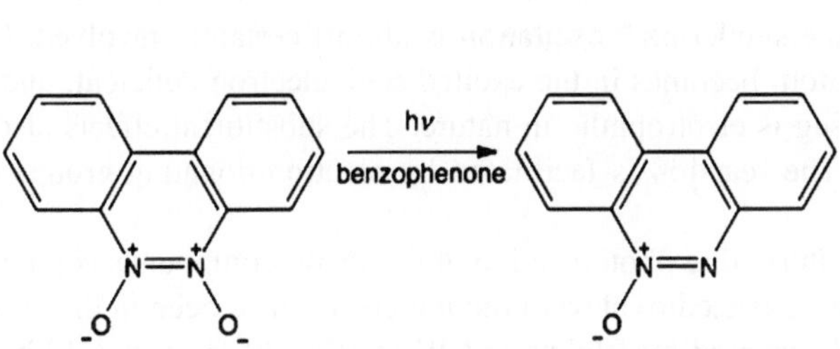

Scheme 3.92

formation of azoxybenzene, two reaction mechanisms were proposed. Mauser and Heitzer [287] proposed that the first reaction step is a recombination of the triplet excited molecules to the dihydroxy intermediate, which by elimination of water molecule yields azoxybenzene (Scheme 3.90).

An alternative mechanism [272] includes reduction of the azodioxide from its triplet state (Scheme 3.91):

This mechanism could be additionally argued by findings that benzo[*o*]cinnoline dioxide reduces to its azoxide form only in the presence of benzophenone (Scheme 3.92):

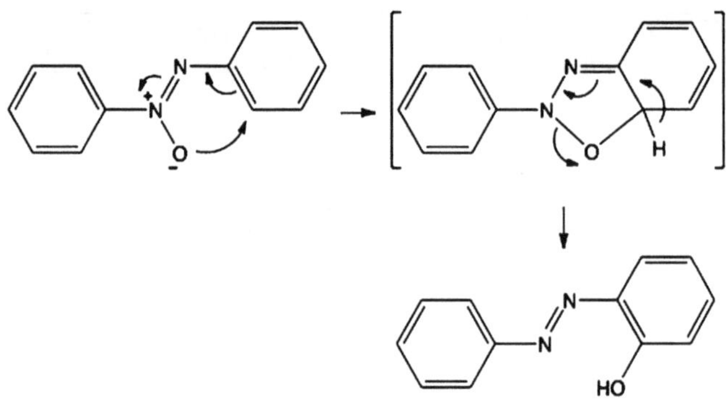

Scheme 3.93

The first proposed mechanisms for the rearrangement of azoxide to 2-hydroxyazobenzene included an intramolecular reaction step in which oxygen nucleophilic attack promotes formation of cyclic transition state [290]. Since five-membered cyclic transition state is sterically favored, the attack of oxygen is preferred to the more distant ring (Scheme 3.93). Experiments with ^{15}N and ^{18}O labeling supported this idea [291, 292].

The photochemical rearrangement mechanism that was also proposed, includes the same type of the reaction intermediate, and involves the singlet $\pi\pi^*$ excited state [293, 294]. However, the detailed investigations by Bunce et al. [295] have shown that the singlet $n\pi^*$ excitation is almost certainly involved. Consequently, the oxygen atom becomes in the excited state electron deficient, and its attack on the distant ring is electrophilic in nature. The substituent effects also support this hypothesis. The reaction is facilitated by electron-donating groups bound to the distant ring.

The most intriguing photoreaction of C-nitroso compounds is photodissociation of nitroso dimers (azodioxides) to monomers, as have been in the case of aliphatic derivatives discovered by Bluhm and Weinstein, 40 years ago [296]. Importance of this reaction for potential practical use is in the evidence that the appearance of monomers change the color of the solution to greenish or blue. Because the difference in the absorption spectra of monomers and dimers is large, and both of the species are persistent in the solution, basic condition for the photochromic effect is satisfied. Nitroso dimer-monomer photochromism of aromatic derivatives has been discovered by Azoulay and Fischer [175], and studied systematically by UV and visible spectroscopy. A series of methyl- and dimethyl substituted nitrosobenzenes were irradiated by UV light in methylene dichloride solution at −60 °C. Since at this temperature the dominant species is dimer, the exposure to the irradiation yields monomers. By warming to the room temperature, and recooling to −60 °C the dimeric forms were recovered. The high efficiency of this photoreaction was confirmed by measured quantum yield, which have values between 1.01 and 1.14.

Scheme 3.94

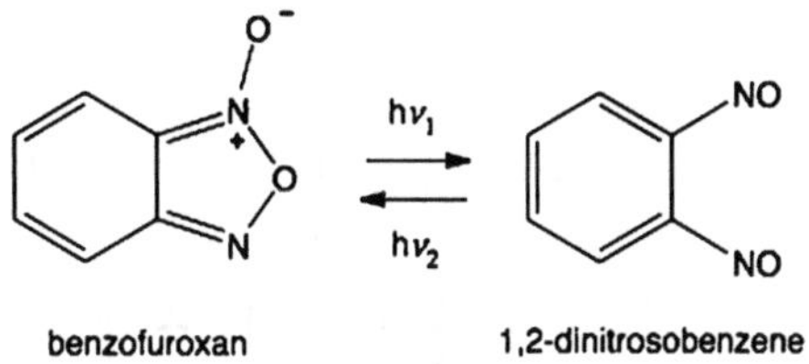

Scheme 3.95

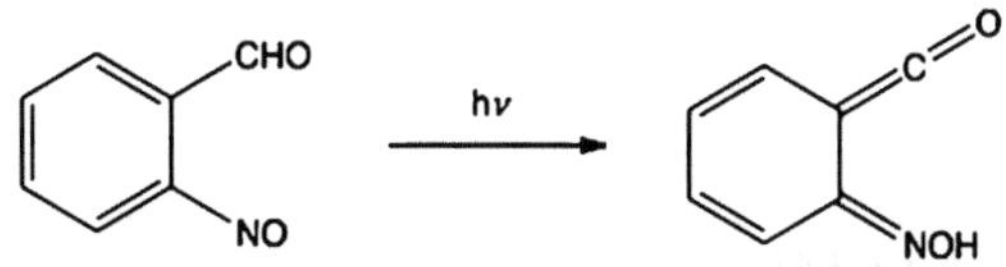

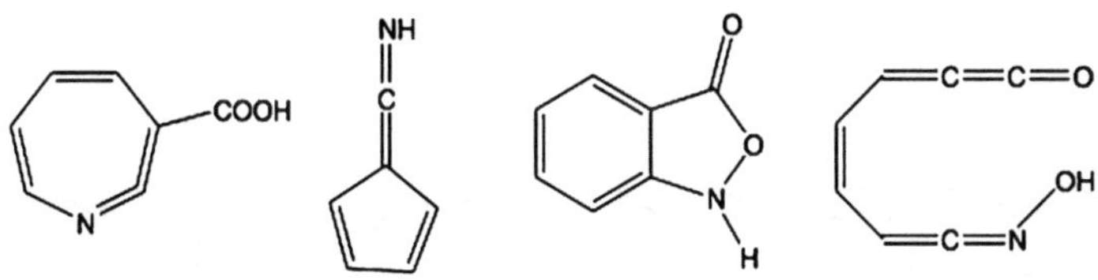

Scheme 3.96

As we have already discussed previously, such a photochromic and thermochromic system is also operative in the solid-state (Sect. 3.3.3).

An interesting photochemical behavior has been observed for 1,2-dinitrosobenzene. This compound really has the structure described as benzofuroxan [297]. Irradiation of benzofuroxan in xenon matrices at 14 K with 366 nm light resulted in the formation of 1,2-dinitrosobenzene (Scheme 3.94) [298].

The resulted dinitrosobenzene was characterized by IR absorptions at 1,515, 1,102, 805, and 795 cm^{-1}. Irradiation of the resulted 1,2-dinitrosobenzene with the 254 or 313 nm light converts it back to benzofuroxane. Interconversion of these species was also confirmed by UV spectroscopy. While benzofuroxan absorbs at 378, 362, and 214 nm, 1,2-dinitrosobenzene has spectral maxima at 280, 266, and 260 nm. Such a photochromic system is efficient, but, unfortunately, operative only at cryogenic temperatures between 12 and 80 K. Further development in experimental [299] and computational studies [300, 301] of the 1,2-dinitrosobenzene-furoxane moiety as well as the parent dinitrosoethylene [302] confirmed that the structures represent only the reactive intermediates.

Photochemistry of substituted nitrosobenzenes could give complicated mixtures of products. For instance, *o*-nitrosobenzaldehyde isolated in argon matrix at 12 K and irradiated with 313 nm UV light undergoes rearrangement to the ketene oxime structure as the main product (Scheme 3.95).

However, other unstable species (Scheme 3.96), listed in the following scheme, have also been detected, and characterized in matrices.

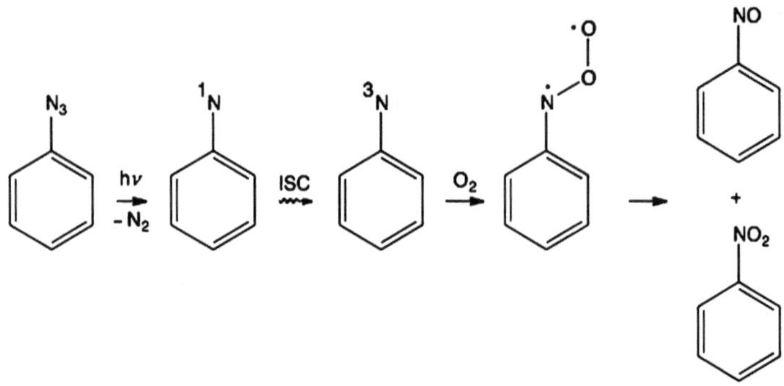

Scheme 3.97

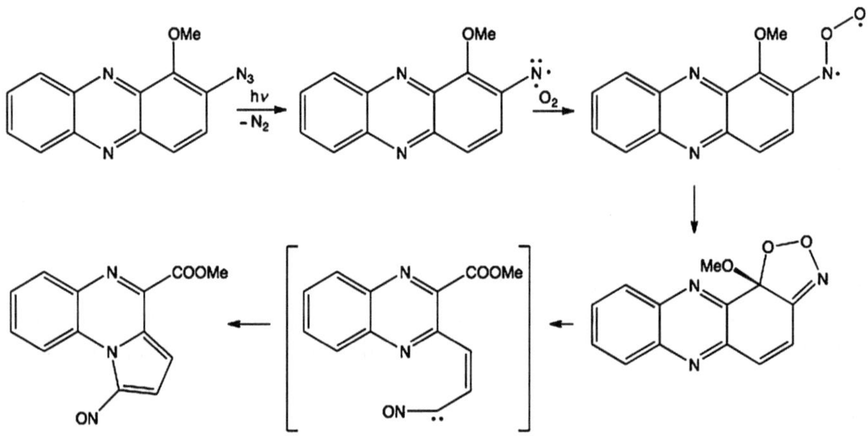

Scheme 3.98

3.5.1 Photoreactions Including Nitrosoaromatic Intermediates

Derivatives of nitrosobenzene appear also as intermediates in a series of photoreactions. Phenylnitrene that is formed after photolysis of phenylazide in low temperature argon matrices can undergo intersystem crossing from singlet to the triplet state. While the singlet phenylnitrene rearranges with the ring extention [303], triplet phenylnitrene could dimerize to azobenzene or react with oxygen to yield nitroso oxide [304–306]. Disproportionation of the nitroso oxide diradical results with the formation of nitrosobenzene and nitrobenzene (Scheme 3.97).

Transformation of aromatic azides to the nitroso compounds has been observed under the preparative experimental conditions already 20 years ago [307] by photooxidation of 2-azido-1-methoxyphenazine (Scheme 3.98). On the basis of high level quantum chemical calculations (UB3LYP/6-311+G(*d,p*) and G3MP2B3), the reaction has later been explained by formation of nitrene that is oxidized to the nitroso oxide [308]. The proposed formation of the carbene (in brackets in the Scheme 3.98) is only tentative.

Aryl nitroso compounds could be also returned to corresponding nitrenes by reaction with trivalent phosphorus compounds (phosphines or phosphites) [309–311].

$$ArNO + R_3P \rightarrow {}^1ArN + R_3PO$$

The reaction can be visualized because the resulting triphenylphosphine oxide affords chemiluminiscence.

References

1. Allan ZJ, Podstata J, Snobl D, Jarkovsky J (1967) Collect Czech Chem Commun 32: 1449–1461
2. (a) Brownstein S, Gabe E, Lee F, Tan L (1984) J Chem Soc Chem Commun:1566–1568; (b) Brownstein S, Gabe E, Lee F, Piotrowski A (1986) Can J Chem 64:1661–1667
3. Laane J, Ohlsen JR (1980) Prog Inorg Chem 27:465–513
4. Keck DB, Hause CD (1968) J Mol Spectrosc 26:163–174
5. Skokov S, Wheeler RA (1999) J Phys Chem 103:4261–4269
6. Reents WD, Freiser BS (1980) J Am Chem Soc 102:271–276
7. Reents WD, Freiser BS (1981) J Am Chem Soc 103:2791–2797
8. Bosch E, Kochi JK (1994) J Org Chem 59:5573–5586
9. Kochi JK (1990) Acta Chem Scand 44:409–432
10. Lindeman SV, Bosch E, Kochi JK (2000) J Chem Soc Perkin Trans 2:1919–1923
11. See any modern textbook of physical organic chemistry
12. Fraser RR, Raganskas AJ, Stothers JB (1982) J Am Chem Soc 104:6475–6476
13. Craik DJ, Levy GC, Brownlee RTC (1983) J Org Chem 48:1601–1606
14. Geske DH, Ragle JL, Bambenek MA, Balch AL (1964) J Am Chem Soc 86:987–1002
15. Hehre WJ, Radom L, PvR S (1986) Ab initio molecular orbital theory. Wiley, London
16. Pross A, Radom L, Taft RW (1980) J Org Chem 45:818–826
17. Exner O, Folli V, Marcaccioli S, Vivarelli P (1983) J Chem Soc Perkin Trans 2:757–760
18. Krygowsky TM, Kalinowsky MK, Turowska-Tyrk I, Hiberty PC, Milart P, Silvestro A, Topsom RD, Daehne S (1991) Struct Chem 2:71–79
19. Irle S, Krygowsky TM, Niu JE, Schwarz WHE (1995) J Org Chem 60:6744–6755
20. Taft WR, Price E, Fox IR, Lewis IC, Andersen KK, Davis GT (1963) J Am Chem Soc 85:709–724; ibidem 3146–3156
21. Exner O, Chapman NB, Shortner J (eds) (1978) Correlation analysis in chemistry. Plenum, New York, p 439, Chapter 10
22. Krygowsky TM, Sadlej-Sosnowska N (2011) Struct Chem 22:17–22
23. Sadlej-Sosnowska N, Krygowsky TM (2009) Chem Phys Lett 476:191–195
24. Fletcher DA, Gowenlock BG, Orrell KG (1998) J Chem Soc Perkin Trans 2:797–803
25. Irle S, Krygowsky TM, Nin JE, Schwarz WHE (1995) J Org Chem 60:6744
26. Fletcher DA, Gowenlock BG, Orrell KG, Šik V, Hibbs DE, Hursthouse MB (1996) J Chem Soc Perkin Trans 2:191
27. Fletcher DA, Gowenlock BG, Orrell KG, Šik V (1995) Magn Reson Chem 33:561
28. Beepath N, Stephenson D (2004) Hyperfine Interact 159:137–141
29. Holschbach MH, Sanz D, Claramunt RM, Infantes L, Motherwell S, Raithby PR, Jimeno ML, Herrero D, Alkorta I, Jagerovic N, Elguero J (2003) J Org Chem 68:8831–8837

30. Lumsden MD, Wu G, Wasylishen RE, Curtis RD (1993) J Am Chem Soc 115:2825–2832
31. Kronja O, Matijević-Sosa J, Uršić S (1987) J Chem Soc Chem Commun:463–464
32. Uršić S (1993) Helv Chem Acta 76:131–138
33. Pilepić V, Lovrek M, Vikić-Topić D, Uršić S (2001) Tetrahedron Lett 42:8519–8522
34. Vinković Vrček I, Pilepić V, Uršić S (2004) Tetrahedron Lett 45:699–702
35. Uršić S, Lovrek M, Vinković Vrček I, Pilepić V (1999) J Chem Soc Perkin Trans 2:1295–1297
36. Lovrek M, Pilepić V, Uršić S (2000) Croat Chem Acta 73:715–731
37. Pilepić V, Jakobušić C, Vikić-Topić D, Uršić S (2006) Tetrahedron Lett 47:371–375
38. Moodie RB, Osullivan B (1995) J Chem Soc Perkin Trans 2:205–208
39. Zuman P, Shah B (1994) Chem Rev 94:1621–1641
40. Kress G, Manthey H (1956) Chem Ber 89:1412–1422
41. Asmus KD, Wigger A, Henglein A (1966) Ber Bunsec-Ges Phys Chem 70:862–868
42. Darchen A, Moinet C (1977) J Electroanal Chem 78:81–88
43. Lotlikar PD, Miller EC, Miller JA, Margreth A (1965) Cancer Res 25:1743–1753
44. Neumann HG, Meltzer M, Topner W (1977) Arch Toxicol 39:21–30
45. Alberti A, Benaglia M, Carloni P, Greci L, Stipa P, Rizzoli C, Sgarabotto P (1995) Gazz Chim Ital 125:555–561
46. Capozzi G, Modena G, Patai S (eds) (1974) The chemistry of the thiol group. Wiley, London
47. Diepold C, Eyer P, Kampffmeyer H, Reinhardt K (1982) Asv Exp Med Biol 136B:1173–1181
48. Montanari S, Paradisi C, Scorrano G (1999) J Org Chem 64:3422–3428
49. Shah B (1993) Dissertation. Clarkson University, Potsdam, NY
50. Bamberger E, Buesdorf H, Szolayski B (1899) Chem Ber 32:210–221
51. Darchen A, Moinet C (1976) J Chem Soc Chem Commun:820a
52. Lo Conte M, Carroll KS (2012) Angew Chem Int Ed 51:6502–6505
53. Ehrlich P, Sachs F (1899) Chem Ber 32:2341–2346
54. Takahashi K, Kimura S, Ogawa Y, Yamada K, Ida H (1978) Synthesis:892–893
55. Osdene TS, Timmis GM (1954) Chem Ind (London):405–406
56. Bell F (1957) J Chem Soc:516–518
57. Anderson DMW, Bell F (1959) J Chem Soc:3708–3713
58. Yang L, Tan B, Wang F, Zhong G (2009) J Org Chem 74:1744–1746
59. Payette JN, Yamamoto H (2008) J Am Chem Soc 130:12276–12278
60. Qin L, Li L, Yi L, Da C-S, Zhou Y-F (2011) Chirality 23:527–533
61. Zhong G, Yu Y (2004) Org Lett 6:1637–1638
62. Hayashi Y, Yamaguchi J, Sumiya T, Hibino K, Shoji M (2004) J Org Chem 69:5966
63. Brown SP, Brochu MP, Sinz CJ, MacMillan DWC (2003) J Am Chem Soc 125:10808
64. Jiao P, Yamamoto H (2009) Synlet 16:2685
65. Jiao P, Kawasaki M, Yamamoto H (2009) Angew Chem Int Ed 48:3333
66. Ramachary DP, Barbas CF III (2005) Org Lett 7:1577, Org Chem 69:5966
67. Trisler JC, Deland PH, Goodgame MM (1975) Tetrahedron Lett 36:3125
68. Sasaki T, Ishibashi Y, Ohno M (1983) Chem Lett 22:863
69. Sasaki T, Mori K, Ohno M (1985) Synthesis 3:279
70. Momiyama N, Yamamoto H (2002) Angew Chem Int Ed Engl 41:2986
71. Momiyama N, Yamamoto H (2002) Angew Chem Int Ed Engl 41:3313
72. Momiyama N, Yamamoto H (2005) Chem Commun:3514
73. Momiyama N, Yamamoto H (2004) J Am Chem Soc 126:6038
74. Momiyama N, Yamamoto H (2004) J Am Chem Soc 126:6498
75. Momiyama N, Yamamoto H (2004) J Am Chem Soc 126:5360
76. Momiyama N, Yamamoto H (2002) Org Lett 4:3579
77. Momiyama N, Yamamoto H (2005) J Am Chem Soc 127:1080
78. Srivastava RS (2003) Tetrahedron Lett 44:3271
79. Zhang FL, Schweizer WB, Xu M, Vasella A (2007) Helv Chim Acta 90:521
80. Wu YM, Ho LY, Cheng CH (1985) J Org Chem 50:392
81. Suzuki K, Weisburger EK (1966) Tetrahedron Lett 44:5409
82. Suzuki K, Weisburger EK (1968) J Chem Soc C:199

83. Yunes RA, Terenzani AJ, Amaral LD (1975) J Am Chem Soc 97:368
84. Ueno K, Akiyoshi S (1954) J Am Chem Soc 76:3670
85. Ogata Y, Takagi Y (1958) J Am Chem Soc 80:3591
86. Yunes RA, Meyer MM, Terenzani AJ, Andrich DD, Scarabino CC (1969) Rev Fac Ign Quim 38:239
87. Yunes RA, Terenzani AJ, Andrich DD, Scarabino CA (1970) Rev Fac Ign Quim 39:181
88. Yunes RA, Terenzani AJ, Andrich DD, Scarabino CA (1971–1972) Rev Fac Ign Quim 40–41:11
89. Yunes RA, Terenzani AJ, Andrich DD, Scarabino CA (1973) J Chem Soc Perkin Trans 2:696
90. Froyen P (1971) Acta Chem Scand 25:2781
91. Merino E (2011) Chem Soc Rev 40:3835
92. Giumanini AG, Toniutti N, Verardo G, Merli M (1999) Eur J Org Chem:141
93. Kyne RE, Ryan MC, Kliman LT, Morken JP (2010) Org Lett 12:3796
94. Olah GA (1995) Angew Chem Int Ed Engl 34:1393
95. Vančik H, Sunko DE (1989) J Am Chem Soc 111:3742
96. Buzek P, Schleyer PR, Vančik H, Mihalić Z, Gauss J (1994) Angew Chem Int Ed Engl 33:448
97. Olah GA, Fung AP, Rawdah TN (1980) J Org Chem 45:4149
98. Olah GA, Kiovsky TE (1968) J Am Chem Soc 90:6461
99. Okamoto T, Shudo K, Ohta T (1975) J Am Chem Soc 97:7184
100. Shudo K, Ohta T, Endo Y, Okamoto T (1977) Tetrahedron Lett:101
101. Olah GA, Donovan DJ (1978) J Org Chem 43:1743
102. Laali KK, Bolvig S, Hansen PE (1995) J Chem Soc Perkin Trans 2:537
103. Dandge DK, Donaruma LG (1987) Nitroso Polymers. In: Kroschwitz JI (ed) Encyclopedia of Polymer Science and Engineering, vol 10. Wiley, New York, p 185
104. Gruger A, Le Calvé N (1975) Spectrochim Acta 31A:581
105. Leach AG, Houk KN (2002) J Chem Soc Chem Commun:1243
106. Shibusaki M, Sodeoka M, Ogawa Y (1984) J Org Chem 49:4098
107. Anson CE, Hartmann S, Kelsey RD, Richard Stephenson G (2000) Polyhedron 19:569
108. Lightfoot AP, Pritchard RG, Wan H, Warren JE, Whiting A (2002) Chem Commun:2072
109. Foldes RS, Vandor K, Kollar L, Horvath J, Tuba Z (1999) J Org Chem 64:5921
110. Templin SS, Wallock NJ, Bennett DW, Siddiquee T, Haworth DT, Donaldson WA (2007) J Heterocyclic Chem 44:719
111. Lemire A, Beaudoin D, Grenon M, Charette AB (2005) J Org Chem 70:2368
112. Anand A, Bhargava G, Singh P, Mahajan MP (2009) Heterocycles 77:547
113. Gamenara D, Días E, Tancredi N, Heinzen H, Moyna P, Forbes EJ (2001) J Braz Chem Soc 12:489
114. Krchnak V, Moellmann U, Dahse HM, Miller MJ (2008) J Comb Chem 10:104
115. Yamamoto Y, Momiyama N, Yamamoto H (2004) J Am Chem Soc 126:5962
116. Zhu R, Zhang D, Wu J, Liu C (2007) J Mol Struct (THEOCHEM) 815:105
117. Buergi HH (1998) Acta Cristallogr A54:873
118. Buergi HH, Dunitz JH (1983) Acc Chem Res 16:153
119. Roth-Barton J, White JM (2009) Aust J Chem 62:1695
120. Penoni A, Palmisano G, Zhao YL, Houk KN, Volkman J, Nicholas KM (2009) J Am Chem Soc 131:653
121. Kang JK, Bugarin A, Connell BT (2008) Chem Commun:3522
122. Xu ZJ, Zhu D, Zeng X, Wang F, Tan B, Hou Y, Lv Y, Zhong G (2010) Chem Commun 46:2504
123. Piloty O (1898) Ber Dt Chem Ges 31:456
124. Piloty O (1901) Ber Dt Chem Ges 34:1863
125. Piloty O (1902) Ber Dtsch Chem Ges 35:3090
126. Lüttke W (1954) J Physique et la Radium 15:633
127. Lüttke W (1956) Angew Chem 67:235
128. Lüttke W (1956) Angew Chem 68:417
129. Lüttke W (1957) Angew Chem 69:99
130. Hammick DL (1931) J Chem Soc:3105

131. Chilton HTJ, Gowenlock BG, Trotman J (1955) Chem Ind:538
132. Gowenlock BG, Trotman J (1955) J Chem Soc:4190
133. Gowenlock BG, Trotman J (1955) J Chem Soc:1670
134. Glaser R, Murmann RK, Barnes CL (1996) J Org Chem 61:1047
135. Milovac S, Šimunić-Mežnarić V, Vančik H, Višnjevac A, Kojić-Prodić B (2001) Acta Cryst E 57:0218
136. Schaper K (2008) Magn Reson Chem 46:1163. Fletcher DA (1996) In: Ph.D. thesis. University of Exeter, UK (See also Ref. d in Chapter 1). Castelleno SM (1967) Tetrahedron Lett:5205
137. Fletcher DA, Gowenlock BG, Orell KG (1997) J Chem Soc Perkin Trans 2:2201
138. Orell KG, Stephenskon D, Velarque JH (1990) J Chem Soc Perkin Trans 2:1297
139. Orrell KG, Šik V, Stephenson D (1987) Magn Reson Chem:1007
140. Al-Tahou BM, Gowenlock BG (1986) Rec Trav Chim Pays-Bas 105:353
141. Cox RH, Hamada M (1979) Org Magn Reson 12:322
142. Grishin YK, Sergeyev NM, Subbotin OA, Ustynuk YA (1973) Mol Phys 25:279
143. Negrebetskii VV, Bokanov AI, Stepanov BI (1981) Zh Strukt Khim 22:88
144. Gowenlock BG, Cameron M, Boyd ASF, Al-Tahou BM, McKenna P (1994) Can J Chem 72:514
145. Salzmann R, Wojdelski M, McMahon M, Havlin RH, Oldfield E (1998) J Am Chem Soc 120:1349
146. Mason J, Lakworthy LF, Moore EA (2002) Chem Rev 102:913
147. Dahn H, Pechy P, Flogel R (1994) Helv Chim Acta 77:306
148. Wu G, Zhu J, Mo X, Wang R, Terskikh V (2010) J Am Chem Soc 132:5143
149. Burawoy A, Cais M, Chamberlain JT, Liversedge F, Thompson AR (1955) J Chem Soc:3721
150. Tabei K, Nagakura S (1965) Bull Chem Soc Jpn 38:965
151. Holleck L, Schindler R (1956) Z Elektrochem 60:1142
152. Jacquemin D, Perpéte EA (2006) Chem Phys Lett 420:529
153. Zharkova OM, Morozova YP, Trubacheva EN, Artyukhov VY (2009) Russ Phya J 52: 458–463
154. Engert JM, Slenczka A, Kensy U, Dick B (1996) J Phys Chem 100:11883
155. Von Keussler V, Lüttke W (1959) Z Elektrokem 63:614
156. Siegbahn K et al (1969) ESCA applied to free molecules. North-Holland, Amsterdam
157. Batich CD, Donald DS (1984) J Am Chem Soc 106:2758
158. Perrine KA, Leftwich TR, Weiland CR, Madachik MR, Opila RL, Teplyakov AV (2009) J Phys Chem 113:6643
159. Liftwich TR, Madachik MR, Teplyakov AV (2008) J Am Chem Soc 130:16216
160. Rademacher P (1996) In: Patai S (ed) Chemistry of amino, nitroso, nitro and related groups. Wiley, Chichester/New York/Brisbane/Toronto/Singapore
161. Bergmann H, Elbel S, Demuth R (1977) J Chem Soc Dalton Trans:401
162. Egdell R, Green JC, Rao CNR, Gowenlock BG, Pfab J (1976) J Chem Soc Faraday Trans 2 72:988
163. Ernsting NP, Pfab J, Green JC, Römelt J (1980) J Chem Soc Faraday Trans 2 76:844
164. Lacombe S, Loudet M, Dargelos A, Camou JM (2000) Chem Phys 258:1
165. Rabalais JW, Colton R (1972/1973) J Electron Spectrosc Relat Phenom 1:83
166. Egdell R, Green JC, Rao CNR (1975) Chem Phys Lett 33:600
167. Stevens WR, Ruscic B, Baer T (2010) J Phys Chem A 114:13134
168. Gowenlock BG, Maidment MJ, Orrell KO, Šik V, Mele G, Vasapollo G, Hursthouse MB, Abdul Malik KM (2000) J Chem Soc Perkin Trans 2:2280–2286
169. Lüttke W (1957) Z Elektrochem 61:976
170. Nakamoto K, Rundle RE (1958) J Am Chem Soc 78:1113
171. Darwin C, Hodgkin DC (1950) Nature 166:827
172. Lüttke W (1957) Z Elektrochem 61:302
173. Wajer TAJW, de Boer TJ (1972) Recueil 91:565
174. Azoulay M, Lippman R, Wettermark G (1981) J Chem Soc Perkin Trans 2:256

175. Azoulay M, Fischer E (1982) J Chem Soc Perkin Trans 2:637
176. Janbon S, Davey RJ, Shankland K (2008) CrystEngComm 10:279
177. Talberg HJ (1979) Acta Chem Scand Ser A 33:289
178. Talberg HJ (1978) Acta Chem Scand Ser A 32:401
179. Talberg HJ (1977) Acta Chem Scand Ser A 31:485
180. Mijs WJ, Hokstra SE, Ullman RH, Havinga E (1958) Rec Trav Chim 77:746
181. Holmes R (1964) J Org Chem 29:3076
182. Holmes RR, Bayer RP, Errede LA, Davis HR, Wiesenfeld AW, Bergman PM, Nicholas DL (1965) J Org Chem 30:3837
183. Semin IV, Sokolenko VA, Tovbis MS (2007) Russ J Org Chem 43:544
184. Alemasov YE, Slaschinin DG, Tovbis MS, Kirik SD (2011) J Mol Struct 985:184
185. Dieterich DA, Paul IC, Curtin DY (1974) J Am Chem Soc 96:6372
186. Azoulay M, Drakenberg T, Wettermark G (1979) J Chem Soc Perkin Trans 2:199
187. Nietzki R, Kehrmann F (1887) Bericht 20:613
188. Gowenlock BG, Richter-Addo GB (2005) Chem Soc Rev 34:797–809
189. Childress WL, Donaruma LG (1974) Macromolecules 7:427
190. Boyer JH, Toggweiler U, Stoner GA (1957) J Am Chem Soc 79:1748
191. Anderson L, Cameron M, Gowenlock BG, McEwen IJ (1992) J Chem Soc Perkin Trans 2:243
192. McKenna P (1986) Ph.D. thesis, Heriot-Watt University
193. Orrell KG, Šik V, Stephenson D (1987) Magn Reson Chem 25:1007
194. Klyuchnikov OR, Chachkov DV, Deberdeev RY, Zaikov GE (2005) Russ J Appl Chem 78:315
195. Hacker NP (1993) Macromolecules 26:5937–5942
196. Mallory FB, Mannatt SL, Wood CS (1965) J Am Chem Soc 87:5433
197. Gowenlock BG, McCullough KJ (1989) J Chem Soc Perkin Trans 2:551
198. Hedayatullah M, Raoult J-C, Denivelle L (1973) Bull Soc Chim Fr:2702
199. Mutlu H, Kusefoglu SH (2009) J Appl Polym Sci 113:1925
200. Zhuzhgov EL, Strunina NV, Komarov VF (1972) Kinet Katal 13:1405
201. Gan LM, Chew CH (1983) Rubber Chem Technol 56:883
202. Klyuchnikov OR, Khairutdinov FG, Klyuchnikov YA (2004) Russ J Appl Chem 77:1382
203. Drefahl G, Horhold H-H, Hofmann KD (1968) J Prakt Chem (Leipzig) 37:91
204. Yoneda A, Sugihara K, Hayashi K, Tanaka M (1973) Kobunshi Kagaku 30:180 (Chem Abstr 1973, 32377a)
205. Mason J, Dunderdale J (1956) J Chem Soc:754
206. Cameron M, Gowenlock BG, Vasapollo G (1991) J Organomet Chem 403:325
207. Gowenlock BG, McCullough KJ (1993) J Chem Res (S):360, J Chem Res (M):2481
208. Schmidt GMJ et al (1976) In: Ginsburg D (ed) Solid state photochemistry. Verlag Chemie, Weinheim/New York, p 80
209. Kim JH, Hubig SM, Lindeman SU, Kochi JK (2001) J Am Chem Soc 123:8795
210. Avrami M (1939) J Chem Phys 7:1103
211. Avrami M (1940) J Chem Phys 8:212
212. Avrami M (1941) J Chem Phys 9:177
213. Erofeev BV (1946) Comput Rend Acad Sci USSR 52:511
214. Hancock JD, Sharp JH (1972) J Am Ceram Soc 55:74
215. Vančik H, Šimunić-Mežnarić V, Meštrović E, Halasz I (2004) J Org Chem 69:4829
216. Vančik H, Šimunić-Mežnarić V, Ćaleta I, Meštrović E, Milovac S, Mlinarić-Majerski K, Veljković J (2002) J Phys Chem B 106:1576
217. Salvalaglio M, Vetter T, Giberti F, Mazzoti M, Parrinello M (2012) J Am Chem Soc 134:17221–17233
218. Halasz I, Meštrović E, Čičak H, Mihalić Z, Vančik H (2005) J Org Chem 70:8461
219. Fletcher DA, Gowenlock BG, Orrell KG, Šik V, Hibbs DE, Hursthouse MB, Malik AKM (1996) J Chem Soc Perkin Trans 2:191
220. Webster MS (1956) J Chem Soc:2841
221. Halasz I, Vančik H (2011) CrystEngComm 13:4307
222. Rassat A, Rey P (1971) J Chem Soc Chem Commun:1161

223. Ullman E, Call L, Tseng SS (1973) J Am Chem Soc 95:1677
224. Ullman EF, Singh P (1972) J Am Chem Soc 94:5077
225. Dunkin IR, Lynch MA, Boulton AJ, Henderson N (1991) J Chem Soc Chem Commun:1178
226. Mišković A, Medančić T, Milovac S, Biljan I, Halasz I, Vančik H (2011) Croat Chem Acta 84:21
227. Biljan I, Kralj M, Mišić Radić T, Svetličić V, Vančik H (2011) J Phys Chem C 115:20267
228. Ross L, Barclay C, Carson DL, Gray JA, Grossman M, Milton JR, Scott CE (1978) Can J Chem 56:2665
229. Batyuk VA, Shabatina TI, Morozov YN, Ryapisov SV, Sergeev GB (1988) Vestn Moscow Univ Seriya 2. Chimia 29:270
230. Halasz I, Biljan I, Novak P, Meštrović E, Plavec J, Mali G, Smrečki V, Vančik H (2009) J Mol Struct 918:19–25
231. Biljan I, Cvjetojević G, Smrečki V, Novak P, Mali G, Plavec J, Babić D, Mihalić Z, Vančik H (2010) J Mol Struct 979:22–26
232. Anandan K, Kolandaivel P, Kumaresan R, Gowenlock BG (2003) J Mol Struct (THEOCHEM) 639:213
233. Anandan K, Kolandaivel P, Kumaresan R, Gowenlock BG (2004) J Mol Struct (THEOCHEM) 680:149
234. Hoffmann R, Gleiter R, Mallory FB (1970) J Am Chem Soc 92:1460
235. Minato T, Yamabe S, Oda H (1982) Can J Chem 60:2740
236. Lüttke W, Skancke PN, Traetteberg M (1994) Theor Chim Acta 87:321
237. Karakus C, Zuman P (1995) J Electroanal Chem 396:499
238. Núñez-Vergara LJ, Santander P, Navarrete-Encina PA, Squella JA (2005) J Electroanal Chem 580:135
239. Núñez-Vergara LJ, Bontá M, Sturm JC, Navarrete PA, Bollo S, Squella JA (2001) J Electroanal Chem 506:48
240. Bollo S, Finger S, Sturm JC, Núñez-Vergara LJ, Squella JA (2007) Electrochim Acta 52:4892
241. Núñez-Vergara LJ, Santander P, Navarrete-Encina PA, Valenzuela J, Sturm JC, Squella JA (2006) J Electrochem Soc 153:E144
242. Lamoureux C, Moinet C, Tallec A (1986) J Appl Electrochem 16:819
243. Lamoureux C, Moinet C (1988) Bull Soc Chim Fr:59
244. Gault C, Mionet C (1989) Tetrahedron 45:3429
245. Guilbaud-Criqui A, Moinet C (1992) Bull Soc Chim Fr 129:295
246. Chuang L, Fried I, Elving PJ (1964) Anal Chem 36:2426
247. Kemula W, Kublik Z (1958) Rocz Chem 32:941
248. Asirvatham MR, Hawley MD (1974) J Electroanal Chem 54:179
249. Laviron E, Vallat A, Meunier-Prest R (1994) J Electroanal Chem 379:427
250. Núñez-Vergara LJ, Squella JA, Olea-Azar C, Bollo S, Navarrete-Encina PA, Sturm JC (2000) Electrochim Acta 45:3555
251. Tian D, Jin B (2011) Electrochem Acta 56:9144
252. Landolt-Barnstein (1989) New series, vol 17, Magnetic properties of free radicals Subvolumes dl and d2. Springer, Berlin
253. Omelka L, Kováčová J (1994) Magn Res Chem 32:525
254. Kaur H, Leung KHW, Perkins MJ (1981) J Chem Soc Chem Commun:142
255. Breza M (2004) J Mol Struct (THEOCHEM) 683:167
256. Pelikán P, Omelka L, Brudíková K, Breza M (2003) J Mol Struct (THEOCHEM) 624:251
257. Ignarro LJ (1989) Circ Res 65:1
258. Davies CA, Nielsen BR, Timmins G, Hamilton L, Brooker A, Guo R, Symons MCR, Winyard PG (2001) Nitric oxide: Biol Chem 5(2):116
259. Guo R, Davies CA, Nielsen BR, Hamilton L, Symons MCR, Winyard PG (2002) Biochim Biophys Acta 1572:133
260. Bard AJ, Gilbert JC, Goodin RD (1974) J Am Chem Soc 96:620
261. Gronchi G, Courbis P, Tordo P, Mousset G, Simonet J (1983) J Phys Chem 87:1343
262. Gronchi G (1993) Res Chem Intermed 19:733
263. McIntire GL, Blount HN, Stronks HJ, Shetty RV, Janzen EG (1980) J Phys Chem 84:916

264. Culcasi M, Tordo P, Gronchi G (1986) J Phys Chem 90:1403
265. Haseloff RF, Mertsch K, Rohde E, Baeger I, Grigor'ev IA, Blasig IE (1997) FEBS Lett 418:73
266. Hiramoto K, Ojima N, Kikugawa K (1997) Free Rad Res 27:409–418
267. Astolfi P, Carloni P, Damiani E, Greci L, Marini M, Rizzoli C, Stipa P (2008) Eur J Org Chem 3279:3285
268. Bamberger E (1902) Ber Dtsch Chem Ges 35:1606
269. Morrison HA (1969) In: Feuer H (ed) The chemistry of the nitro and nitroso groups. Interscience, New York, Chap. 4
270. Maassen JA, de Boer TJ (1973) Recl Trav Chim Pays-Bas Belg 92:185
271. Gowenlock BG, Pfab J, Kresze G (1975) Justus Liebigs Ann Chem 1903
272. Tanikaga R (1969) Bull Chem Soc Jpn 42:210
273. Chatgilialoglu C, Ingold KU (1981) J Am Chem Soc 103:4833–4837
274. McMillen DF, Golden DM (1982) Ann Rev Phys Chem 33:439
275. Choo KY, Golden DM, Benson SW (1975) Int J Chem Kinet 7:713
276. Park J, Dyakov IV, Mebel AM, Lin MC (1997) J Phys Chem A 101:6043
277. Ayscough PB, Sealy RC, Woods DE (1971) J Phys Chem 75:3454
278. Keßler A, Slenczka A, Seiler R, Dick B (2001) Phys Chem Chem Phys 3:2819–2830
279. Engert JM, Dick B (1996) Appl Phys B 63:531–535
280. Hatton WG, Hacker NP, Kasai PH (1990) J Chem Soc Chem Commun:227
281. Bartz JA, Everhart SC, Cline JI (2010) J Chem Phys 132:074310
282. Toniolo A, Persico M (2001) J Chem Phys 115:1817
283. Noble M, Qian CXW, Reisler H, Wittig C (1986) J Chem Phys 85:5763
284. Uberna R, Hinchliffe RD, Cline JI (1995) J Chem Phys 103:7934
285. Uberna R, Hinchliffe RD, Cline JI (1996) J Chem Phys 105:9847
286. Tomer JL, Wall MC, Reid BP, Cline JI (1995) J Chem Phys 102:6100
287. Mauser H, Heitzer H (1965) Z Naturforschg 20b:200–203
288. Pak K, Testa AC (1971) J Phys Chem 76:1087
289. Pak K, Testa AC (1981) J Photochem 16:223–230
290. Badger GM, Buttery RG (1954) J Chem Soc:2243
291. Shemyakin MM, Maimind VI, Vaichunaite BK (1960) Izv Akad Nauk SSSR Ser Khim 866 (Bull Acad Sci USSR Div Chem Sci (1960)) 808
292. Oae S, Fukumoto T, Yamagami M (1963) Bull Chem Soc Jpn 36:601
293. Tanikaga R, Maruyama K, Goto R, Kaji A (1966) Tetrahedron Lett:5925
294. Tanikaga R (1968) Bull Chem Soc Jpn 41:1664
295. Bunce NJ, Schoch JP, Zerner MC (1977) J Am Chem Soc 99:7986–7991
296. Bluhm AL, Weinstein J (1967) Nature 215:1478–1479
297. Boulton AJ, Gosh PB (1969) In: Katritzky AR, Boulton AJ (eds) Advances in heterocyclic chemistry, vol 10. Academic, New York
298. Hacker NP (1991) J Org Chem 56:5216–5217
299. Vichard D, Halle JC, Huguet B, Pouet MJ, Riou D, Terrier F (1998) Chem Commun:791
300. Ponder M, Fowler JE, Schaefer HF (1994) J Org Chem 59:6431
301. Friedrichsen W (1994) J Phys Chem 98:12933. See also Rauhut G (1996) J Comput Chem 17:1848
302. Stevens J, Schweizer M, Rauhut G (2001) J Am Chem Soc 123:7326–7333
303. Borden WT, Gritsan NP, Hadad CM, Karney WL, Kemnitz CR, Platz MS (2000) Acc Chem Res 33:765
304. Liang TY, Schuster GB (1987) J Am Chem Soc 109:7803
305. Gritsan, NP (2007) Usp Khim 76:1218, Russ Chem Rev (2007) 76:1139
306. Pritchina EA, Gritsan NP, Bally T (2006) Phys Chem Chem Phys:719
307. Albini A, Bettinetti G, Minoli G (1987) J Org Chem 52:1245
308. Talipov MR, Khursan SL, Safiullin RL (2012) Russ J Phys Chem A 86:235–243
309. Chainikova EM, Safiullin RL (2009) Russ Chem Bull Int Ed 58:926–928
310. Cadogan JIG (1968) Q Rev Chem Soc 22:222
311. Chainikova EM, Teregulova AN, Shamukaev VA, Safiullin RL (2009) High Energy Chem 43:147–148

Chapter 4
Organometallic Compounds

Organometallic chemistry of the nitrosoaromatic compounds covers a very large variety of the structures. The nitroso group is responsible for the formation of differently structured complexes with a large group of metals. In first part of this chapter, the main structural types of complexes are analyzed from the structural and electronic aspects. Behavior of such complexes as catalysts for syntheses of complex organic structural motifs is also commented. In the second section the formation of complexes with the metaloporphirines as model systems for the chemistry of the reactions of the heme-containing molecules is discussed in more details.

4.1 Structures and Properties of Nitrosoaromatic-Metal Complexes

Since the aromatic nitroso molecule possesses π-bond and two hetero-atoms with lone electron pairs, its behavior as a ligand affords a huge diversity. Arylnitroso molecule can coordinate with N-atoms, O-atoms and the π–NO bond. In principle, eight coordination modes (Scheme 4.1) of monomeric C-nitroso compounds to metal were proposed [1].

In the first group (Type I–III) the N- O-, or π-nitroso ligand is coordinated only to one metal atom. In the second group (Type IV–VIII) are coordination structures with two or more metal atoms. Metal atoms could be various, from iron, which appears with carbonyl ligands, to ruthenium, platinum, and palladium.

If nitrosoaryl molecule in *ortho* position has a substituent which can coordinate with the metal atom, the additional types (Scheme 4.2) of complexes were found [2].

Cameron and Gowenlock [3] have systematized different types of aromatic nitroso-metal complexes in the accordance with the periodic classification (Table 4.1). The aromatic nitroso ligands, which were used for such a classification, include nitrosobenzene- (**NbX**) or the nitrosodimethylaniline derivatives (**NbD**).

H. Vančik, *Aromatic C-nitroso Compounds*, DOI 10.1007/978-94-007-6337-1_4,

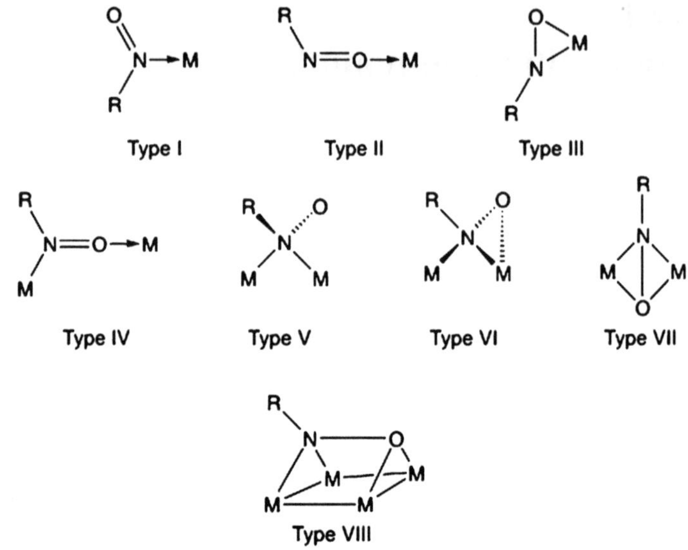

Scheme 4.1

Scheme 4.2

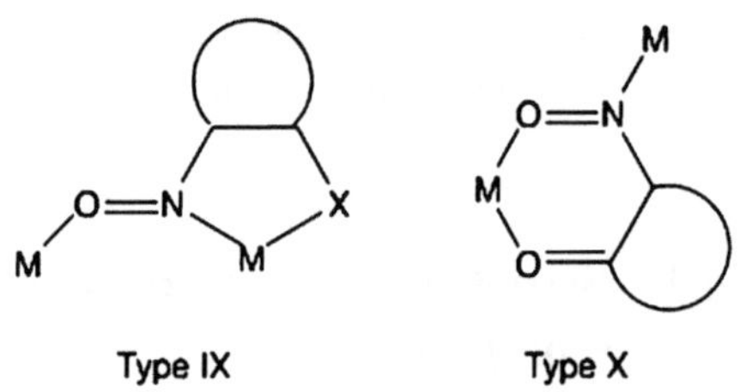

The first crystal and molecular structure of nitrosoaromatic iron tricarbonyl complexes has been obtained for bis[tricarbonyl-(3-chloro-2-methylnitrosobenzene)-iron] (Scheme 4.3) [4].

From the measured N–O distance (1.40 Å) that is closer to the single than to the double bond, the complex could belong to the Type IV mode, where the coordination is obtained by σ_N and π_{NO} orbitals. Additionally, such a lengthening can be explained by the electron back donation from the iron *d*-orbitals into the π^*_{NO}.

Interest in investigation of such nitrosoarene-metal compounds has been grown after discoveries that introduction of the ArN unit changes catalytic properties of transition metal complexes [5]. Moreover, it has been demonstrated that nitrosoarene complexes appear as intermediates in the metal carbonyl and the cyclopentadienylmetal deoxygenation of nitroarenes [6]. For instance, allylic amination by nitroaryl derivatives is catalyzed by dinuclear complexes $[Cp^{\wedge}M(CO)_2]_2$ ($Cp^{\wedge} = C_5H_5$, M = Fe, Ru): [7]

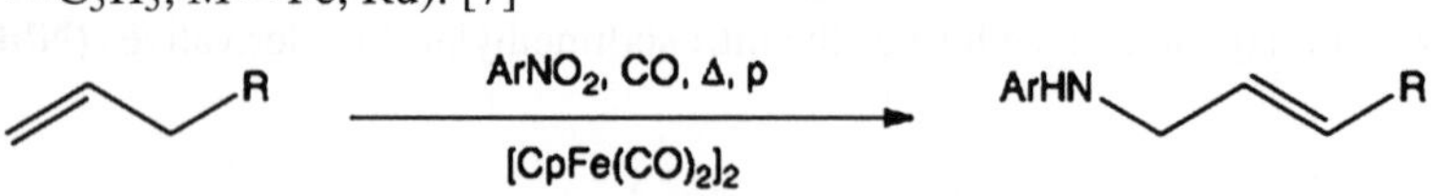

Table 4.1 Systematization of the organometallic complex types found for the C-nitroso compounds (Adopted from Ref. [3])

3	4	6	7	8	9	10	11	12	14
	Ti	**Cr**	**Mn**	**Fe**	**Co**	**Ni**	**Cu**	**Zn**	**Ge**
	–	I	I	I	I	–	–	II	
	NbX	NbX	NbX	NbX	NbD	NbD	NbD	NbD	
				VI	VI	III	II		
				NbX	NbX	NbX	NbD		
						NbD			
					IX				
					NbX				
					X				
					NbX				
		Mo		**Ru**	**Rh**	**Pd**		**Cd**	**Sn**
		I		I	I	I		–	II
		NbX		NbX	NbX	NbX		NbX	NbD
				NbD	NbD	NbD			
		III			VI	III		II	–
		NbX			NbX	NbX		NbD	
						VI			
						NbX			
La		**W**	**Re**	**Os**	**Ir**	**Pt**		**Hg**	
–		I	I	III	I	I		II	
NbD		NbX	NbX	NbX	NbX	NbX		NbD	
					NbD				
					II				
					NbD				
		III	VII			III			
		NbX	NbX			NbX			
						NbD			
						IV			
						NbX			
						V			
						NbX			

Scheme 4.3

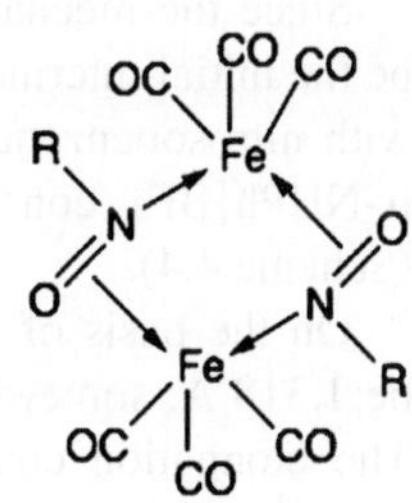

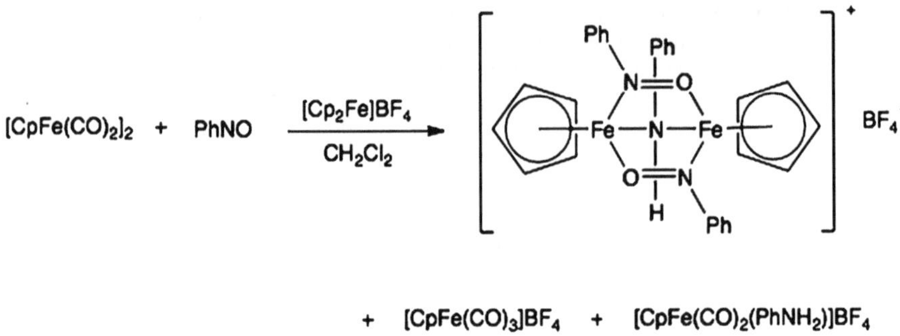

Scheme 4.4

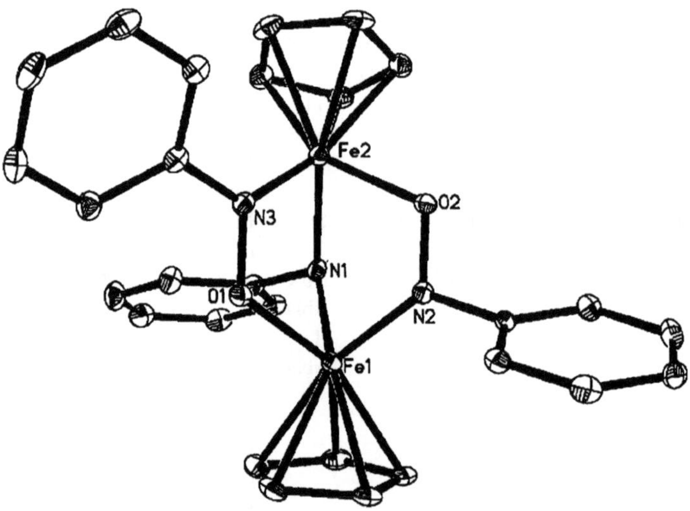

Fig. 4.1 X-ray ORTEP diagram of the cation of the complex described in Scheme 4.4 (Reproduced by permission from Ref. [5])

Since the mechanism includes reduction of nitroarene, nitroso derivative must be the initial intermediate. By ferrocenium salt catalyzed reaction of $[CpFe(CO)_2]_2$ with nitrosobenzene, the stable dinuclear complex $\{[CpFe\text{-}\mu\text{-}(\eta^2\text{-}(N,O)\text{-}PhNO)]_2\text{-}\mu\text{-}NHPh\}BF_4$ consisting from two nitrosobenzene ligands has been isolated (Scheme 4.4).

On the basis of X-ray structural analysis, the N–O distances are measured to be 1.318 Å, somewhat longer then in the free nitrosobenzene monomer (1.25 Å). The elongation could be explained by the electron back donation from Fe to the π^*-NO orbital. The source of the hydrogen atom in the PhNH ligand is unknown, but could origin from the solvent. The molecular structure is shown in the Fig. 4.1.

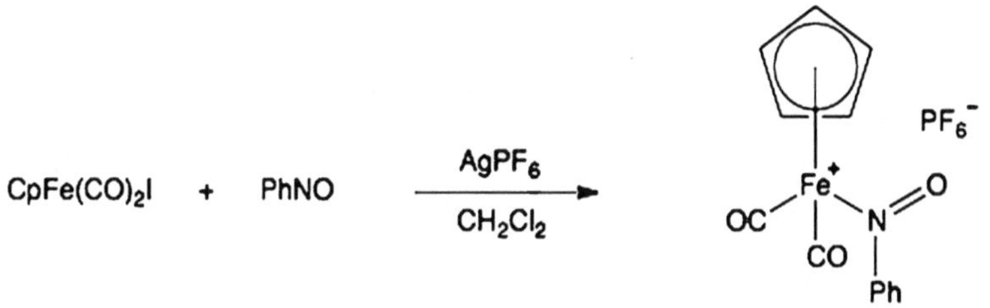

Scheme 4.5

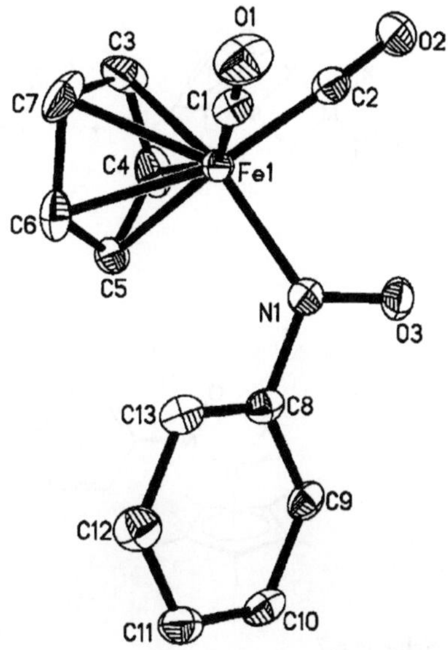

Fig. 4.2 X-ray ORTEP diagram of the cation of the complex described in Scheme 4.5 (Reproduced by permission from Ref. [5])

The cyclopentadienecarbonyliron complex with nitrosobenzene ligand, identified as $[CpFe(CO)_2(PhNO)]PF_6$ (Scheme 4.5) could be prepared by iodide abstraction from $CpFe(CO)_2I$ in the presence of $AgPF_6$.

Its structure is represented in the Fig. 4.2.

The structure belongs to the Type I coordination mode that is common to a series of metal-nitrosoligand complexes [8]. The N–O interatomic distance of 1.226 Å, is shorter in comparison with the 1.25 Å value in the parent ligand. Interesting property of this complex is its possible electrophilicity that follows from the positive charge on the nitroso N-atom (+1.0, based on the PM3(TM)-level of theory).

Under high pressure of carbon monoxide, the complex serves as a donor of the nitrosobenzene unit for the Diels-Alder reaction with the diene (Scheme 4.6).

An interesting Type I complexes are formed if dinitrosoaromatic molecules serve as the ligands. In the reaction of 1,4-bis(4-nitrosophenyl)piperazine (BNPP) with iridium and rhodium complexes $[(\eta^5\text{-}C_5Me_5)MX_2]_2$ (M = Ir, Rh; X = halogen) (Scheme 4.7). All the complexes, as well as the ligand BNPP have been characterized by X-ray analysis of the crystal structure [9, 10].

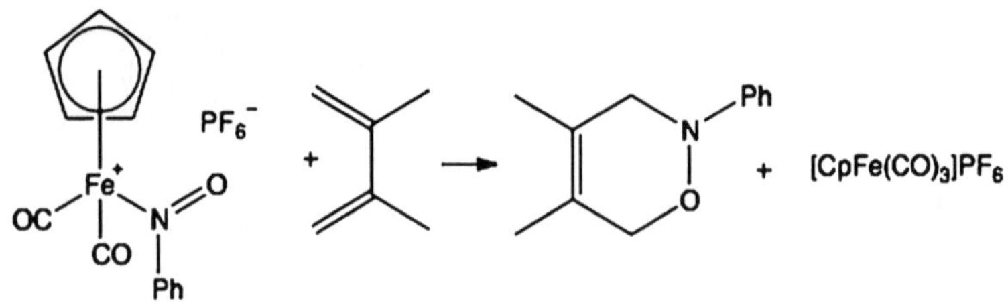

Scheme 4.6

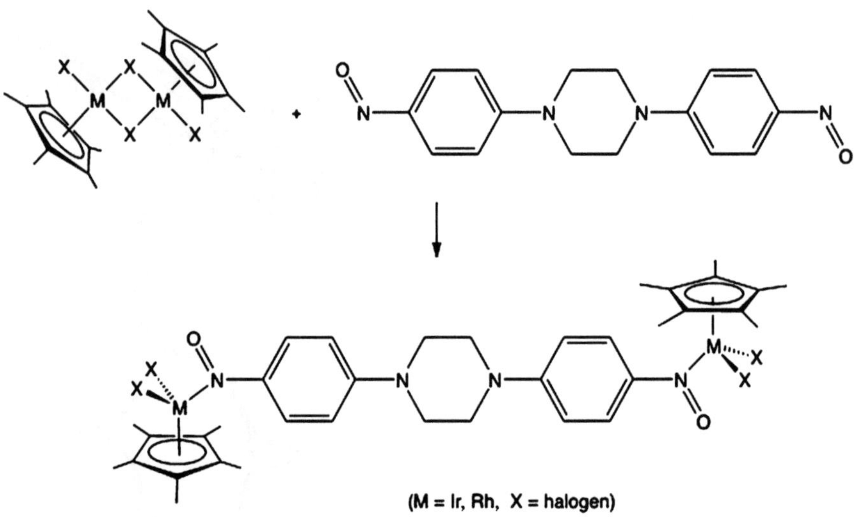

Scheme 4.7

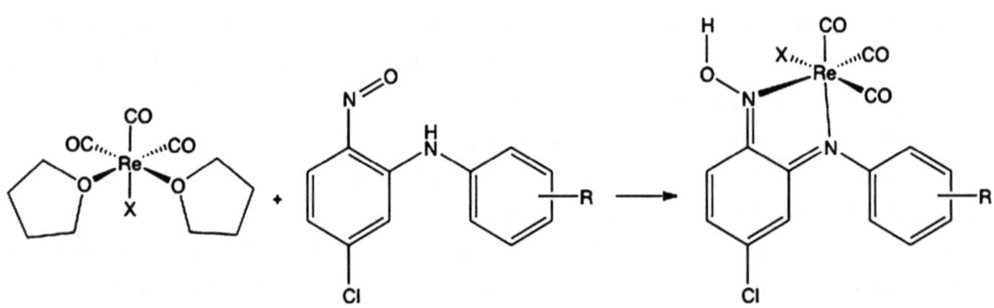

Scheme 4.8

N-coordination can be followed with tautomerization (Scheme 4.8), as it was observed in the reaction of 2-nitroso-N-anilines with rhenium (I). The resulting *N,N′*-chelate complexes are promising substances which could be used as additives in chemotherapy of melanome cancer [11, 12].

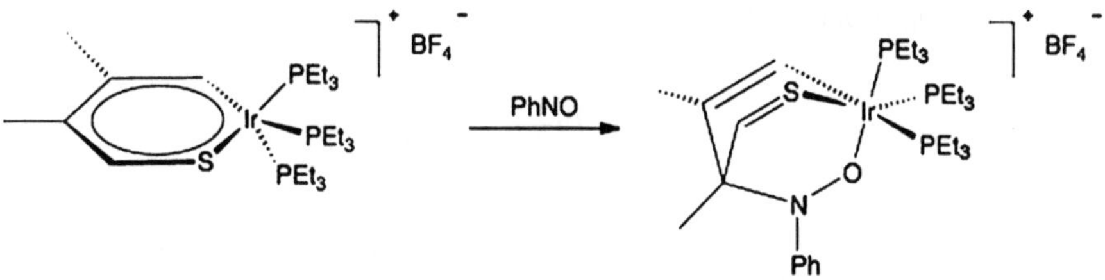

Scheme 4.9

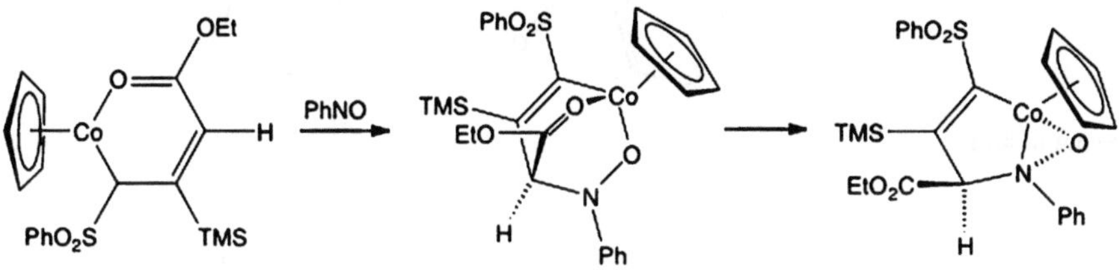

Scheme 4.10

Type II complexes are relatively rare. An interesting example is the cycloaddition product of the reaction of nitrosobenzene with the iridathiabenzene derivative (Scheme 4.9) [13].

The analogous $4\pi + 2\pi$ cycloaddition of nitrosobenzene to the metal complex with the O-atom oriented to the metal center appears also as a first step in the proposed mechanism of the formation of the cobaltocene complex (Scheme 4.10) [14].

Palladium complexes with nitrosoarenes [15, 16] appear as nitroso intermediates during the transition-metal catalyzed reductive carbonylation of nitro arenes [17]. The role of nitrosoarene is in its reaction with the transition-metal carbonyl intermediate, which yields nitrene ligand: [18–20]

$$[M{-}CO] + ONAr \longrightarrow \underset{\text{nitrene ligand}}{[M{=}NAr]} + CO_2$$

Formation of the phenylnitrene as a short living intermediate could be suggested from the structure of the product of the reaction of $Pd_4(\mu\text{-}CO)_4(\mu\text{-}OCOCH_3)_4$ with nitrosobenzene. The phenylnitrene unit in the product complex appears as a substituent in the *ortho* position of the nitrosobenzene ligand (Scheme 4.11). The structure could be classified as the Type IX complex.

The mechanism included in the formation of this specific nitroso ligand could include insertion of the nitrene intermediate to the C–H bond of the nitrosobenzene molecule (Scheme 4.12) [21].

Type IX complexes are also formed in the reaction of rhenium halogenated carbonyls with 1-nitroso-2-naphtol under mild conditions in dichloromethane (Scheme 4.13) [22].

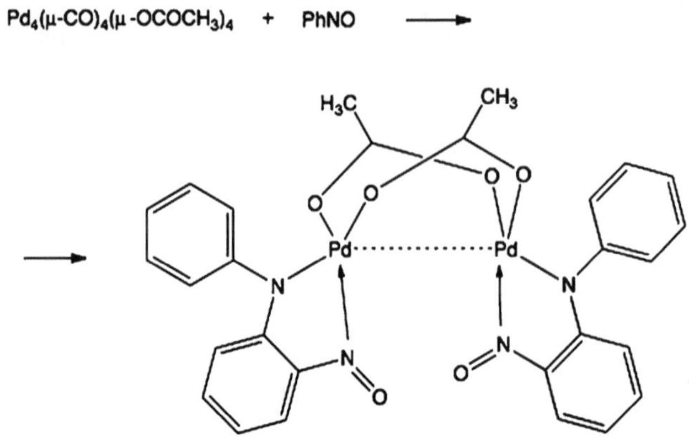

Scheme 4.11

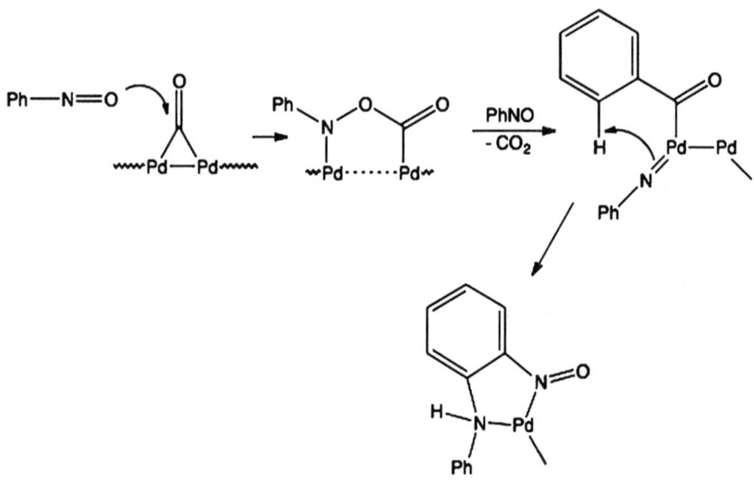

Scheme 4.12

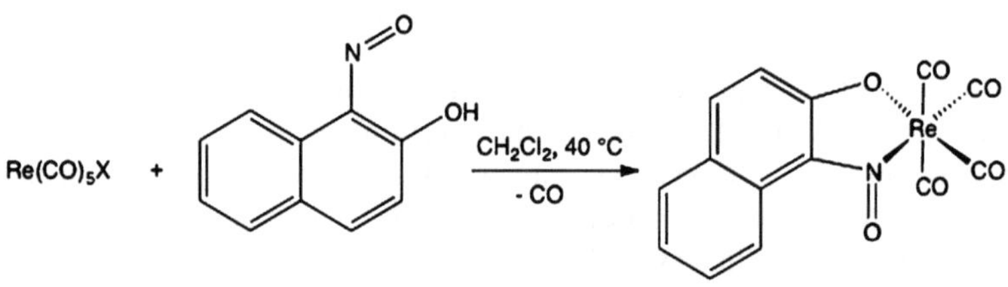

Scheme 4.13

At higher temperatures, the product is formed after the rearrangement of the nitrosonaphtole molecule (Scheme 4.14):

The rearrangement of nitroso to the imino ligand (Scheme 4.15) includes tautomerism followed with reduction.

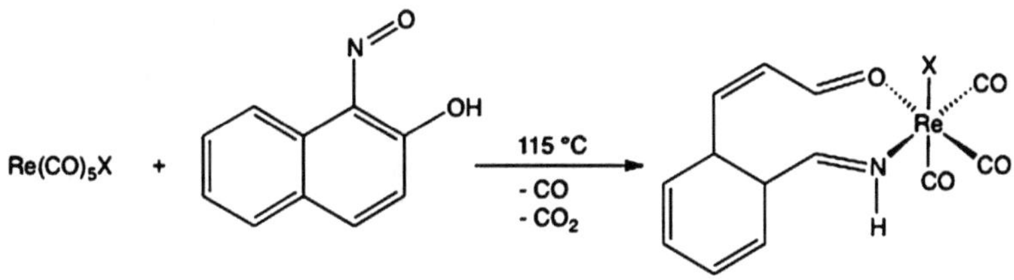

Scheme 4.14

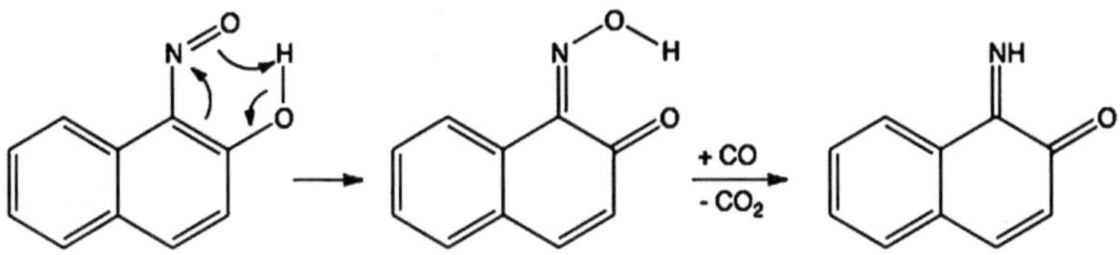

Scheme 4.15

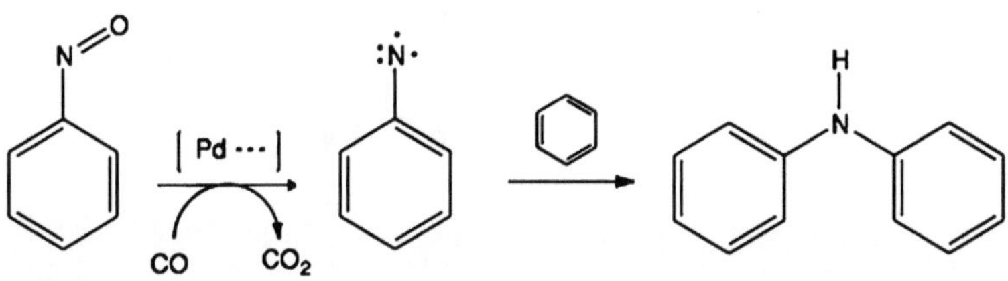

Scheme 4.16

Additional proof for the nitrosobenzene-phenylnitrene transformation is the appearance of the secondary amine byproduct that could be formed by the nitrene insertion in the C–H bond (Scheme 4.16).

Complexes of redox-active metal ions can oxidize nitric oxide, NO, to the active NO^+ ion. In such a way they play active role in nitrosylation of alcohols, amines and thiols [23]. Activation of NO has been observed in reactions of NO with nitrosoaryl complexes of nickel and copper [24]. Such complexes are additionally important because they belong to the Type VII and Type III complexes in which π-electrons of the NO bond are delocalized to the metal atom. Structural consequences of such an interaction are lengthening of the NO bond, and the decrease in the NO stretching frequency. In Type VII complex (Scheme 4.17), $\{[Me_2NN]Ni\}_2(\mu\text{-}\eta^2\text{:}\eta^2\text{-ONAr})$ (Ar = 3,5-dimethylnitrosobenzene) the NO bond length measured by single crystal X-ray diffraction is 1.440 Å, what represents the drastic extension in comparison with the average value obtained for the free nitrosobenzene (1.22 Å). Such geometry change is also followed with the pronounced down-frequency shift of the NO stretching from 1,506 cm^{-1} in free nitrosobenzene molecule, to 915 cm^{-1} in the complex.

Scheme 4.17

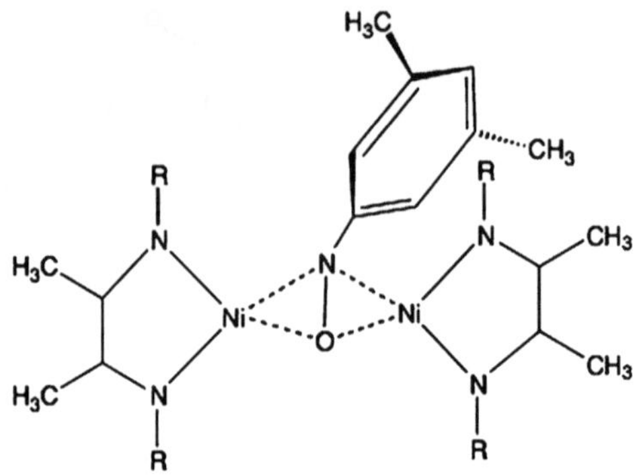

Scheme 4.18

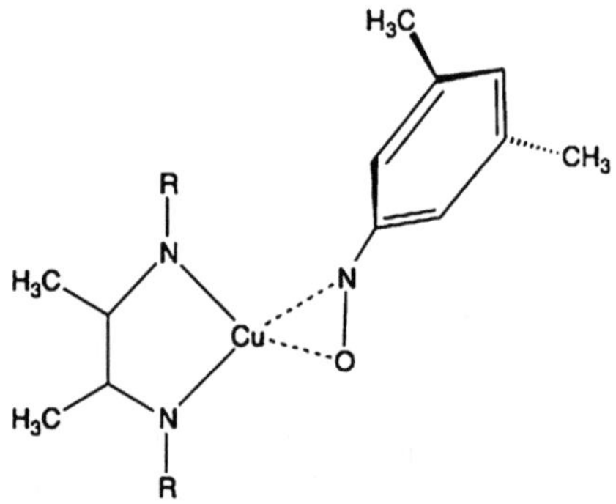

Scheme 4.19

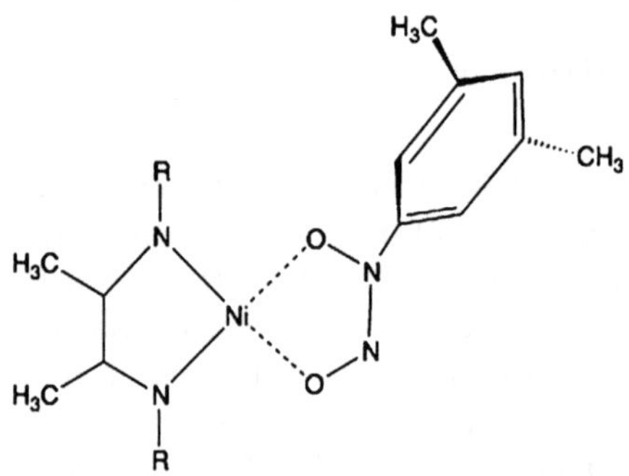

If nickel is replaced with copper, the Type III complex can be prepared [25]. Here the lengthening of the NO bond to 1.333 Å is not so drastic because the π-electron donation from the NO group is only to one metal atom. The NO stretching frequency is lowered to 1,040 cm^{-1}.

Such copper and nickel nitrosoaryl complexes (Schemes 4.18 and 4.19) can assist in activation of the NO molecule [25]. For instance, reaction of the copper complex (represented on the Scheme 4.18) with gaseous NO, led to the formation of the NO adduct [24].

Scheme 4.20

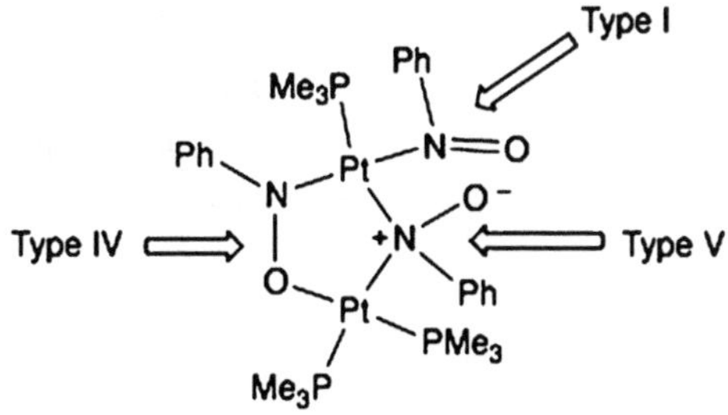

Scheme 4.21

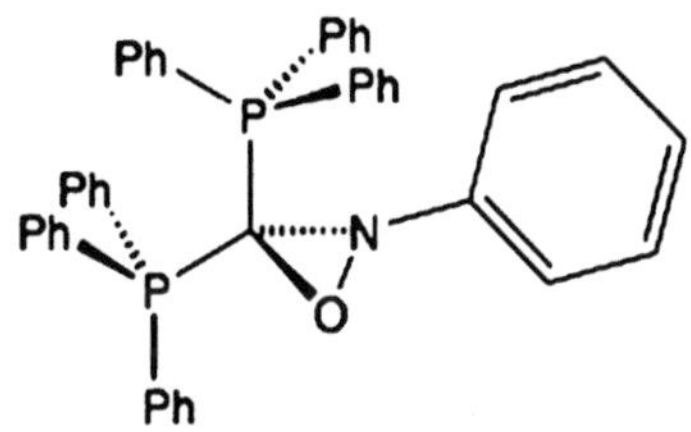

Highly activated NO bond is also found in platinum complex (μ-η^2-nitrosobenzene-*N*)(m-η^1,η^1-nitrosobenzene-*N*,*O*)(η^1-nitrosobenzene-*N*)tris(trimethylphosphine)diplatinum-(II) in which three nitrosobenzene units are bound in a different way [26], Type I, Type IV, as well as the Type V (Scheme 4.20). The extent of the NO bond activation could be estimated from the bond lengths, which have been measured as 1.428 Å for the Type V ligand, and 1.433 Å for the Type IV. In the Type IV adduct, the bond lengthening in the relation to the reference value found in the free nitrosobenzene is as large as 0.20 Å.

Platinum Type III complexes of nitrosobenzene are known with the triphenylphosphine ligands (Scheme 4.21). The platinum atom is positioned in the center of the deformed tetrahedron. Longer NO bond, 1.410 Å, reveals the relatively high extent of NO activation [27].

At room temperature and higher pressure of CO_2, the carbon dioxide molecule can be inserted in the complex by formation of the five-membered ring [28, 29].

Activation of the nitrogen-oxygen bond by the platinum-germylene complex led to the highest NO lengthening known in the literature. In the complex $(Et_3)_2$(PhON)PtGe$[N(SiMe_3)_2]_2$, the NO bond, measured by the single crystal X-ray analysis reach the value of 1.498 Å. The complex, associated to the structural Type IV can be prepared under the mild condition (Scheme 4.22) [30].

Reaction of ruthenium hydrido complex with nitrosobenzene gives the complex Type VI (Scheme 4.23) [31].

In EPR spectrum the complex showed triplet signal with hyperfine splitting with the nitroso nitrogen atom ($g_{iso} = 1.998$, $a_{iso} = 14.2$ G). This observation implies that the unpaired electron is located on the nitrosobenzene ligand. The NO bond (1.353 Å) is elongated relatively to the parent nitrosobenzene.

Scheme 4.22

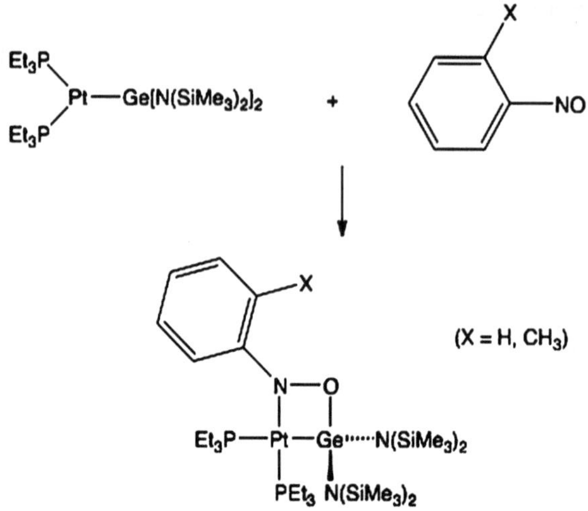

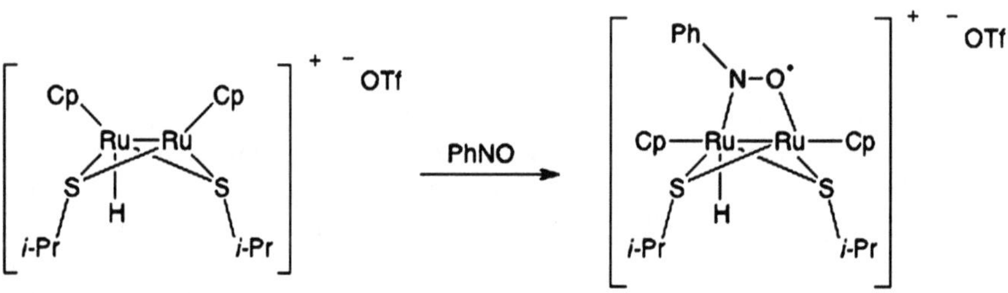

(Cp = cyclopentadienyl, OTf = triflate)

Scheme 4.23

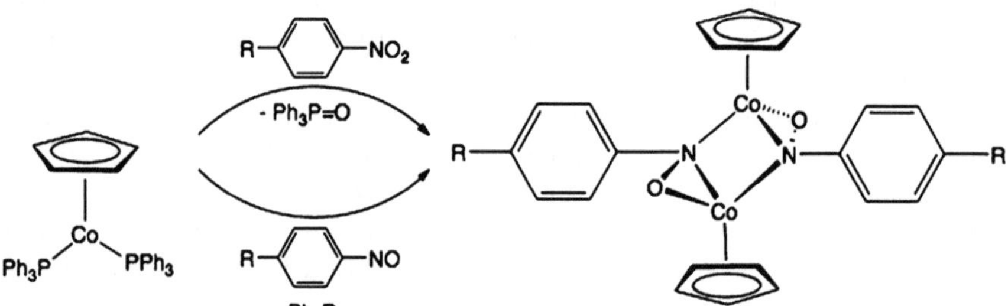

Scheme 4.24

Another example of the Type VI structure is the complex prepared from the cobalt bis(phosphine) cyclopentadiene with arylnitroso or arylnitro compounds [32]. The room temperature reaction led to formation of the dinuclear complex $[(\eta^5\text{-}C_5H_5)Co]_2[\mu\text{-}\eta^1(N)\text{:}\eta^2(N,O)\text{-}N(O)C_6H_5]_2$ (Scheme 4.24).

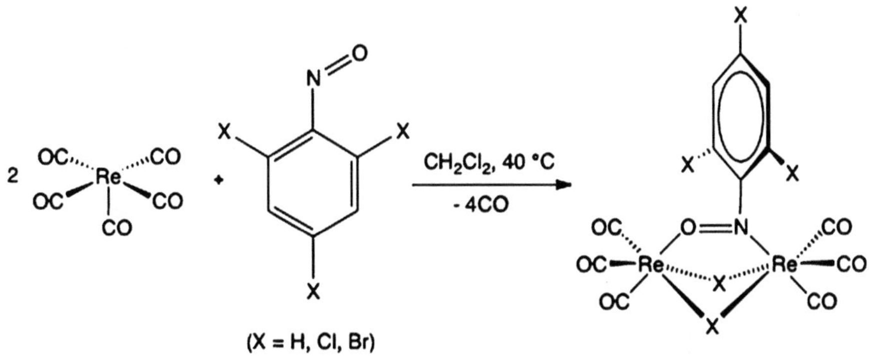

Scheme 4.25

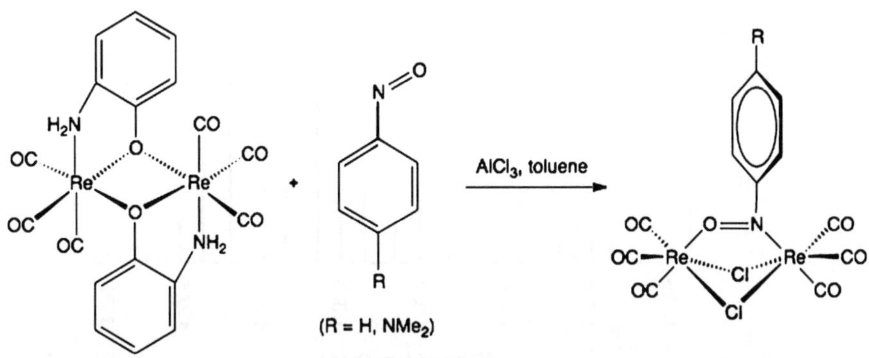

Scheme 4.26

Type VII complexes can be made by reaction of rhenium pentacarbonyl halogenides with 2,4,6-halosubstituted nitrosobenzene derivatives (Scheme 4.25) [2].

Dependently on the halogen substituent, the NO bonds fall in the range 1.257–1.299 Å, slightly longer then in free nitrosobenzene derivatives (1.22 Å). Analogous Type VII complex has been prepared from the reaction of dirhenium complex with nitrosobenzene or *N,N*-dimethylaminonitrosobenzene and $AlCl_3$ in molar ratio 1:1:2 in boiled toluene (Scheme 4.26) [33]. In this compound ($R = NMe_2$) the NO distance is 1.319 Å, and significantly longer than in the free *N,N*-dimethyl-4-nitrosobenzene (1.212 Å).

A series of structurally analogous derivatives with different halogens and R_2N group afford interesting non-linear optical properties. In the dibromo compound with $R = NEt_2$, the dichroic ratio (quotient of intensities of spectral maxima between perpendicular and parallel position of polarizer in relation to the selected crystal orientation) has its maximum at 660 nm [34].

In the niobocene complex Type III (Scheme 4.27) [35], the pronounced activation of the NO bond can only be estimated from the IR spectral data. The NO stretching absorption band appears in the region 1,045–1,097 cm^{-1}, significantly lower than for the free ligand.

Scheme 4.27

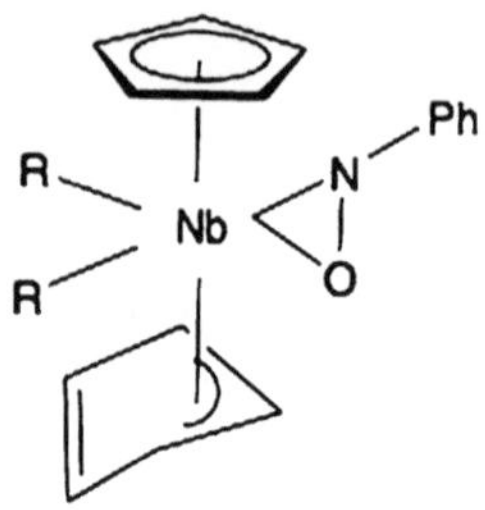

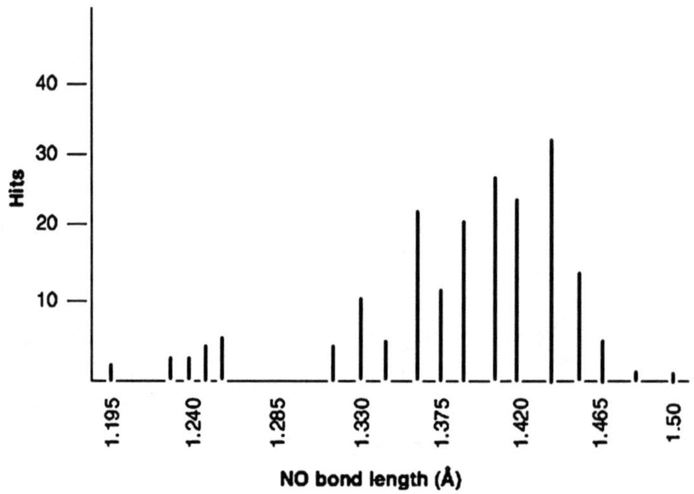

Fig. 4.3 Histogram of NO bond lengths based on the Cambridge Structural Database (*CSD*) (Modified from Ref. [30])

Appearance of N–O bond lengths in the different N-bound nitrosoaryl metal complexes is illustrated on the histogram compiled from a search of the Cambridge Structural Database (CSD) (Fig. 4.3).

4.2 Nitrosoaromatic-Heme Complexes

One of important properties of nitrosoaromatic molecules is their ability of binding to the metal atom of the heme-containing biomolecules [36] such as hemoglobin, myoglobin, cytochrome P450, guanylyl cyclase, and NO synthase. Such a binding of the RNO group could cause inhibition of the biomolecule activity. It has been known for many years that the poisoning activity of nitrobenzene [37, 38] is due to reduction of $PhNO_2$ to PhNOH, which binds in form of PhNO to metallohemoglobin [39–41]. Binding of nitrosobenzene to hemoglobine and myoglobin is even stronger

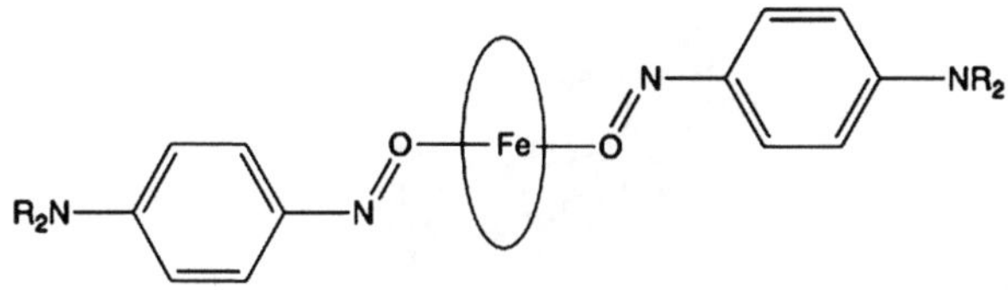

Scheme 4.28

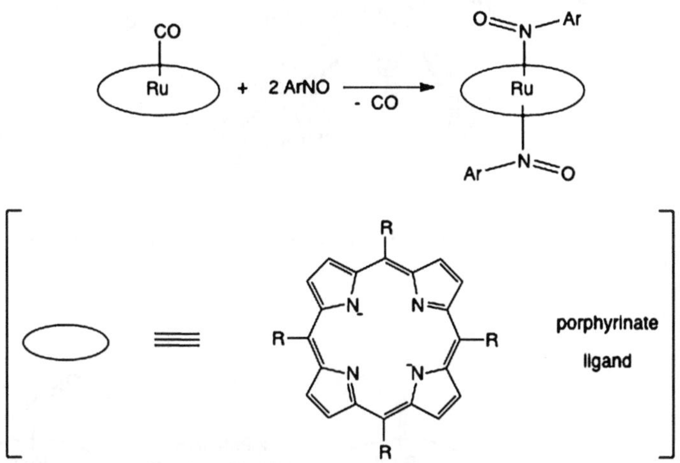

Scheme 4.29

than binding of CO molecule. From that reason, interaction of nitrosobenzene with iron-porphyrins serves as an analog model for investigation of the Fe–O_2 binding in the real biological systems.

Model compounds, which mimic such a behavior of nitroso derivatives in biological systems include RNO complexes with the synthetic metalloporphyrins with different metals (Fe [42], Mn [43], Co [44], Ru [44], Os [45]). First observation of the binding of nitroso compound to the ferrous porphyrine has been made 30 years ago [46] with isopropylnitroso ligand, which binds via the *N*-binding mode.

Nature of binding, *N*-mode (Type I) *versus* *O*-mode (Type II), depends on the oxidation state of the metal [47]. While ferrous complexes prefer an *N*-binding mode, ferric complexes revealed an *O*-binding mode (Scheme 4.28). An example for the *O*-binding complex is $[(5,10,15,20\text{-tetraphenylporphyrin})Fe(ONAr)_2]^+$, in which the Ar group is *p*-aminosubstituted nitrosobenzene. Amino group in *para* position as an *n*-electron donor enhances the negative charge on the NO oxygen atom, what makes *O*-complexing preferable.

Regarding the number of the arylnitroso ligands, there are two types of structures, bis-nitrosobenzene complexes [44, 48] and mono-nitrosobenzene complexes. Bis-nitrosoaryl complexes of rutheniumporphyrin can be prepared from the corresponding carbonyl in the reaction with arylnitroso compound (Scheme 4.29).

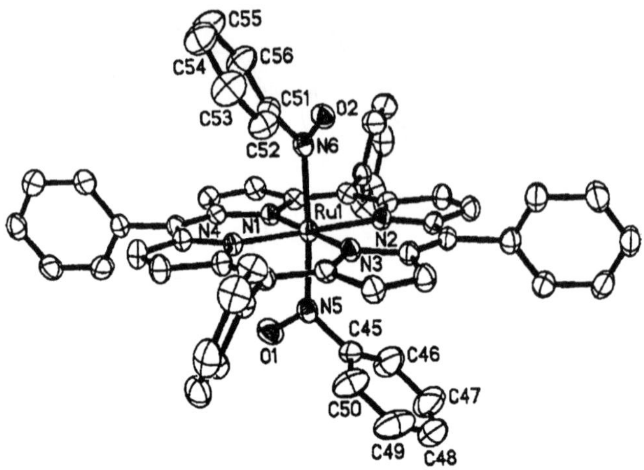

Fig. 4.4 Molecular structure of the complex from the Scheme 4.29 (Reproduced by permission from Ref. [45])

Scheme 4.30

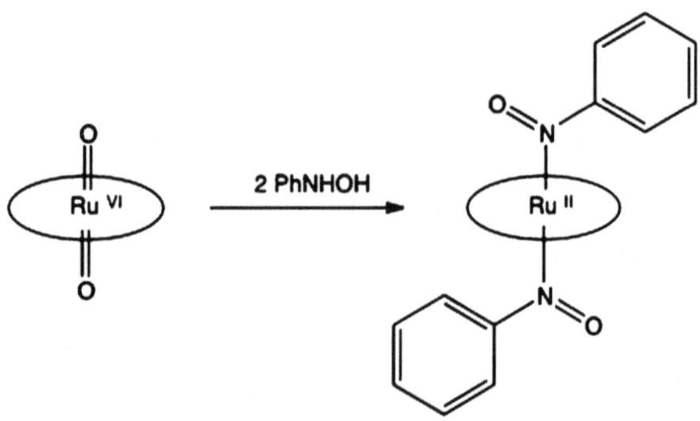

As it is shown in the Fig. 4.4, two nitrosobenzene ligands are axially *N*-coordinated to the ruthenium atom [45].

Reaction of dioxoruthenium(VI) porphyrins with *N*-phenylhydroxylamine in chloroform led to the reduction of the hydroxylamine to the nitrosobenzene, which immediately binds to the metal atom (Scheme 4.30) [49].

However, the reaction products depend strongly on the nature of the porphyrin ring. If the phenyl substituents R on the porphyrin component (*vide supra*) are replaced with 4-methoxyphenyl groups, the reaction with dioxoruthenium(VI) porphyrin led to the disproportionation and formation of the mixed nitrosobenzene-aminobenzene complex (Scheme 4.31).

Osmium analogs (Scheme 4.32) were obtained from the reaction of (por)Os(CO) with PhNO: [50]

Mono-nitrosobenzene complexes are interesting because they are appropriate models for imitating the natural complexes of biomolecules containing porphyrin moiety [42]. In hemoglobin, metal atom centered in the planar porphyrin ring is

Scheme 4.31

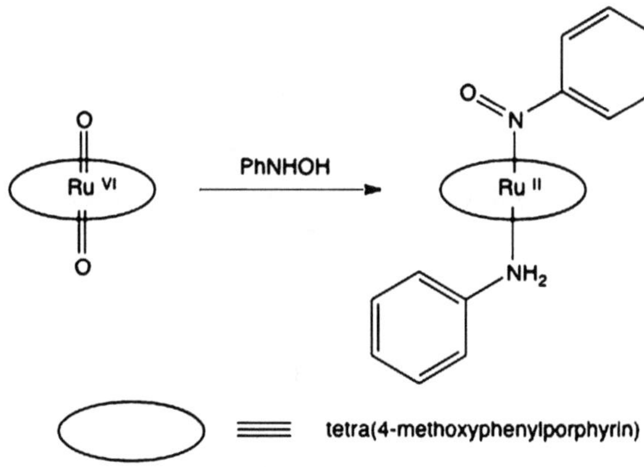

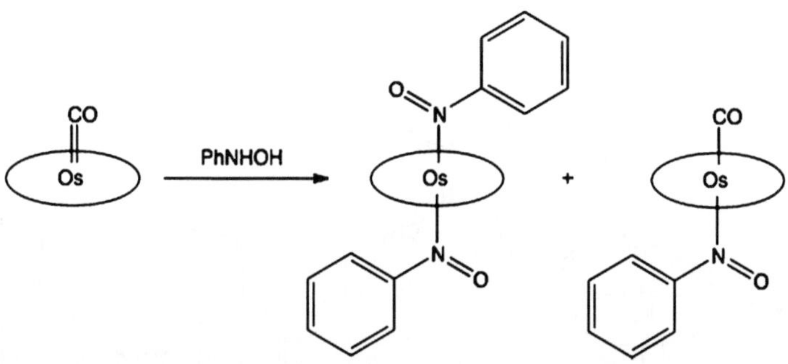

Scheme 4.32

from one side bound to the O_2 molecule, and from the opposite side is complexed with the histidine residue. In the analogous model studies of the iron-O_2 complexing problem, histidine has been replaced either with pyridine or with imidazole ligand (Scheme 4.33).

In Fe(5,10,15,20-tetraphenylporphirinate)-(nitrosobenzene)(pyridine), the PhNO ligand is bound to the iron atom as a complex Type I, with the N–Fe distance 1.818 Å. The CN bond in the nitrosobenzene unit is found to be 1.468 Å, slightly longer than in the free ligand. However, the NO bond length, 1.252 Å is in the complex elongated for 0.03 Å. As it could be seen on the Scheme, replacing the counter-ligand pyridine with 1-methylimidazole does not have larger influence on these geometrical parameters, but the CN bond distance is decreased to 1.444 Å.

In the case of the 4-nitroso-N,N-dimethylaniline ligand, the shortening of the CN bond to 1.437 Å reflects the contribution of the quinoide-like resonance structure (Scheme 4.34).

Model compound with ruthenium porphyrin and *N*-methylimidazol as an axial counter-ligand has been prepared starting with the bis-nitroso complex in reaction with imidazole derivative (Scheme 4.35) [45].

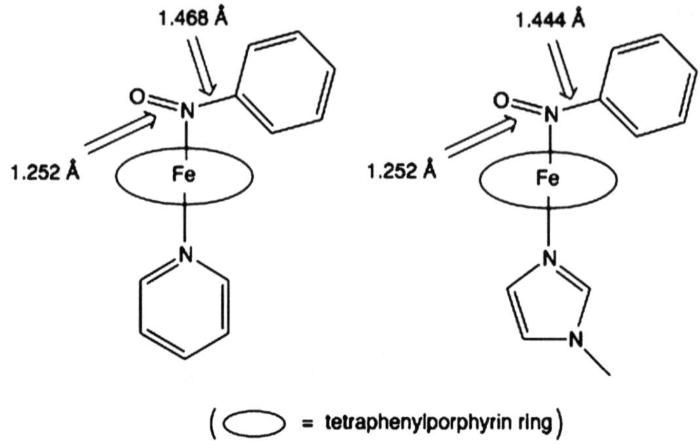

Scheme 4.33

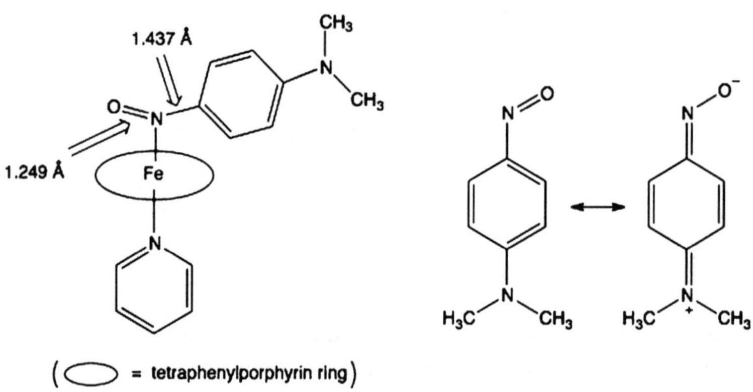

Scheme 4.34

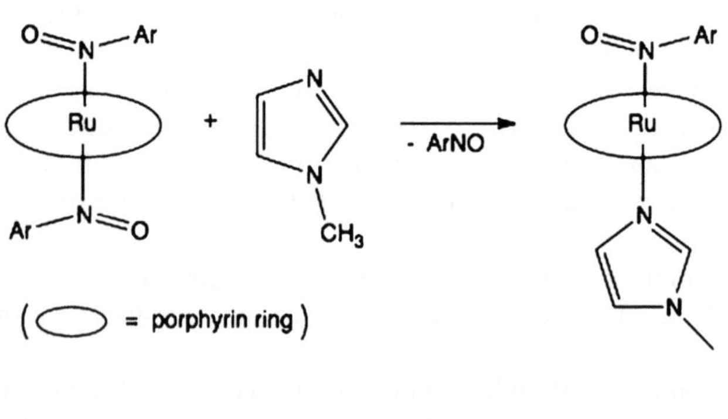

Scheme 4.35

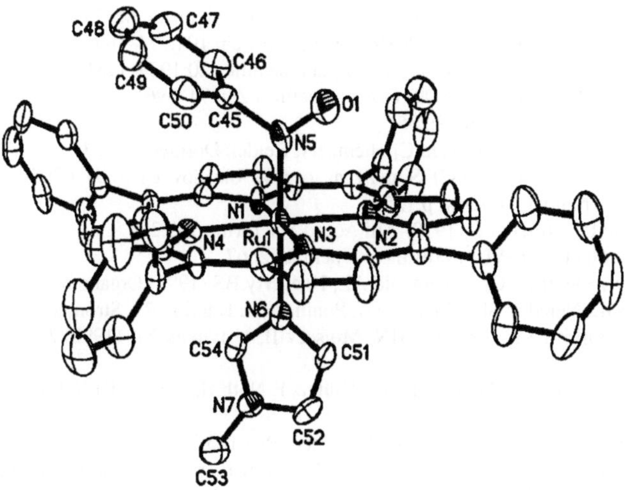

Fig. 4.5 Molecular structure of the complex of nitrosobenzene and *N*-methylimidazol as axial ligands on the porphyrin ring (Reproduced by permission from Ref. [45])

From X-ray structure determination is clear that nitroso component is also *N*-coordinated (Fig. 4.5).

The N–O bond lengths differ in mono- and bis-complexes. In bis-compound, the N–O distances 1.248 and 1.243 Å, respectively, are almost the same as in the parent nitrosobenzene. Donation of electron density from the porphyrin-Ru fragment to NO group (its π^*-orbital) is more evident from the down frequency shift of the NO stretching vibration from 1,506 cm^{-1} in nitrosobenzene to 1,348 cm^{-1} in the complex. As it could be expected, in the mono-complex this shift is even larger: the NO stretching vibration is observed at 1,321 cm^{-1}.

References

1. Cameron M, Gowenlock BG, Parish RV, Vasapollo (1994) Organomet Chem 482:227–240
2. Krinninger C, Wirth S, Klüfers P, Mayer P, Lorenz IP (2006) Eur J Inorg Chem 1060–1066
3. Cameron M, Gowenlock BG (1990) Chem Soc Rev 19:355–379
4. Barrow MJ, Mills OS (1971) J Chem Soc A 864
5. Stephens JC, Khan MA, Nicholas KM (2005) J Organomet Chem 690:4727–4733
6. Berman RS, Kochi JI (1980) Inorg Chem 19:248
7. Srivastava RS, Nicholas KM (1998) Chem Commun 2705
8. Pilato RS, McGettigan C, Geoffroy GL, Rheingold AL, Geib SJ (1990) Organometallics 9:312
9. Alkorta I, Garcia-Gómez C, de Paz JLG, jimeno ML, Arán VJ (1996) J Chem Soc Perkin Trans 2:293
10. Wirth S, Barth F, Lorenz I-P (2012) Dalton Trans 41:2176–2186
11. Wrobel Z, Kwast A (2007) Synlett 1525–1528

12. Wirth S, Wallek AU, Zernickel A, Feil F, Sztiller-Sikorska M, Lesiak-Mieczkowska K, Bräuchle C, Lorenz IP, Czyz M (2010) J Inorg Biochem 104:774–789
13. Bleeke JR, Hinkle PV, Rath NP (2001) Organometallics 20:1939–1951
14. Holland RL, O'Connor JM (2009) Organometallics 28:394–396
15. Moiseev II (1989) Usp Khim 8:2351
16. Keim W (ed) (1983) Catalysis in C_1 chemistry. Reidel, Dordrecht, p 235
17. Yan YB, Nefedov BK (1987) Sintezy na osnove oksidov ugleroda (Carbon Oxide-based syntheses). Khimiya, Moscow, p 125
18. Stanghellini PL, Rossetti R (1970) Atti Accad Sci Torino 105:391
19. Smieja JA, Gladfelter WL (1986) Inorg Chem 25:2667
20. Ramage DL, Geoffroy GL, Rheingold AL, Haggerty BS (1992) Organometallics 11:1242
21. Eremenko IL, Nefedov SE, Sidorov AA, Ponina MO, Danilov PV, Stromnova TA, Stolarov IP, Katser SB, Orlova ST, Vargaftik MN, Moiseev II, Ustynyuk YA (1998) J Organomet Chem 551:171–194
22. Krinninger C, Wirth S, Galvez Ruiz JC, Klüfers P, Nöth H, Lorenz IP (2005) Eur J Inorg Chem 4094–4098
23. Ford PC, Fernandez BO, Lim MD (2005) Chem Rev 105:2439–2456
24. Wiese S, Kapoor P, Williams KD, Warren TH (2009) J Am Chem Soc 131:18105–18111
25. Puiu SC, Warren TH (2003) Organometallics 22:3974–3976
26. Packet DL, Trogler WC, Rheingold AL (1987) Inorg Chem 26:4309
27. Pizzotti M, Porta F, Cenini S, Demartin F, Masciocchi N (1987) J Organomet Chem 330:265–278
28. Bellon PL, Cenini S, Demartin, Pizzotti FM, Porta F (1982) J Chem Soc Chem Commun 265–266
29. Demartin F, Pizzotti M, Porta F, Cenini S (1987) J Chem Soc Dalton Trans 605–608
30. Litz KE, Kampf JW, Banaszak Holl MM (1998) J Am Chem Soc 120:7484–7492
31. Iwasa T, Shimada H, Takami A, Matsuzaka H, Ishii Y, Hidai M (1999) Inorg Chem 38:2851–28591
32. O'Connor JM, Bunker KD (2003) Organometallics 22:5268–5273
33. Wilberger R, Krinninger C, Piotrowski H, Mayer P, Lorenz IP (2004) Eur J Inorg Chem 2488–2492
34. Krinninger C, Högg C, Nöth H, Gálvez Ruiz JC, Mayer P, Burkacky O, Zumbusch A, Lorenz IP (2005) Chem Eur J 11:7228–7236
35. Lucas D, Mugnier Y, Antiñolo A, Otero A, Garcia-Yuste S, Fajardo M (1995) J Organomet Chem 490:7–10
36. Lee J, Chen L, West AH, Richter-Addo GB (2002) Chem Rev 102:1019
37. Filehne W (1898) Arch Exptl Pathol Pharmakol 9:329
38. Loeb RF, Bock AV, Fitz R (1921) Am J Med Sci 539:539
39. Keilin D, Hartree EF (1943) Nature 151:390–391
40. Murayama M (1960) J Biol Chem 235:1024–1028
41. Gibson QH (1960) Biochem J 77:519–526
42. Godbout N, Sanders LK, Salzmann R, Havlin RH, Wojdelski M, Oldfield E (1999) J Am Chem Soc 121:3829
43. Fox SJ, Chen L, Khan MA, Richter-Addo GB (1997) Inorg Chem 36:6465
44. Chen L, Fox JB Jr, Yi GB, Khan MA, Richter-Addo GB (2001) J Porphyrins Phthalocyanines 5:702
45. Lee J, Twamley B, Richter-Addo GB (2002) Can J Chem 80:1252–1258
46. Mansuy D, Battioni P, Chottard JC, Riche C, Chiaroni A (1983) J Am Chem Soc 105:455
47. Wang LS, Chen L, Khan MA, Richter-Addo GB (1996) Chem Commun 323–324
48. Crotti C, Sishta C, Pacheco A, James BR (1988) Inorg Chim Acta 141:13
49. Liang JL, Huang JS, Zhou ZY, Cheung KK, Che CM (2001) Chem Eur J 7:2306–2317
50. Chen L, Khan MA, Richter-Addo GB, Young VG Jr, Powell DR (1998) Inorg Chem 37:4689–4696

Chapter 5
Biological Systems

Behavior of nitrosoaromatic compounds in biological systems is of high scientific interest, because these chemical species, as relatively unstable, could easily react with different biomolecules attacking the regular biochemical processes. As xenobiotic agents, arylnitroso derivatives can display mutagenic [1, 2], carcinogenic [3, 4], as well as cytotoxic [5, 6] effects. Since their oxidation state lies in the middle between the stable oxidized form (nitro derivative) and reduced form (amino or hydroxylamino), nitroso compounds may have marked influence on the oxidations and reductions in living cells. As we know from the photochemistry and electrochemistry represented in previous chapters, nitrosoaromatic molecules readily form radicals and radical ions. Such property is especially important for their reactivity in complex biochemical reaction systems. Ability to behave as ligands with more than ten coordination modes in complexes with broad selection of metal atoms makes these molecules very active in reactions with biogenic metals. Biomolecules, which have active centers in which metal ions are included, could be target for the nitrosoaryl ligands. Many examples of interactions of the heme-constituting molecules with nitrosobenzene were discussed in the previous chapter.

Three classes of reactions of nitrosophenyl compounds with components of living systems were identified, (i) reactions with simpler metabolites such as fatty acids, NAD(P)H, ascorbate, (ii) reactions with thiol functionalized biomolecules such as glutathione, hemoglobine (with its cysteine terminus), myoglobine, and (iii) impact on the redox systems that include radical species that could cause the DNA damage.

5.1 Reactions with Fatty Acids and Steroids

The fate of the nitrosoaryl molecule *in vivo* has been extensively investigated by EPR spectroscopy. After injection of nitrosobenzene in mouse, the two main molecular species were detected, phenylhydronitroxide radical and the adduct with unsaturated fatty acid molecules (Scheme 5.1) [7].

H. Vančik, *Aromatic C-nitroso Compounds*, DOI 10.1007/978-94-007-6337-1_5,

Scheme 5.1

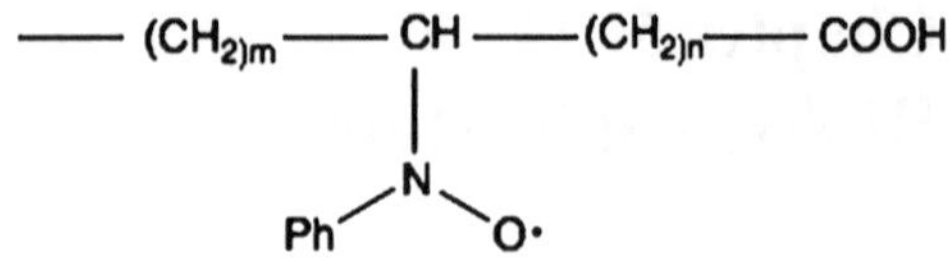

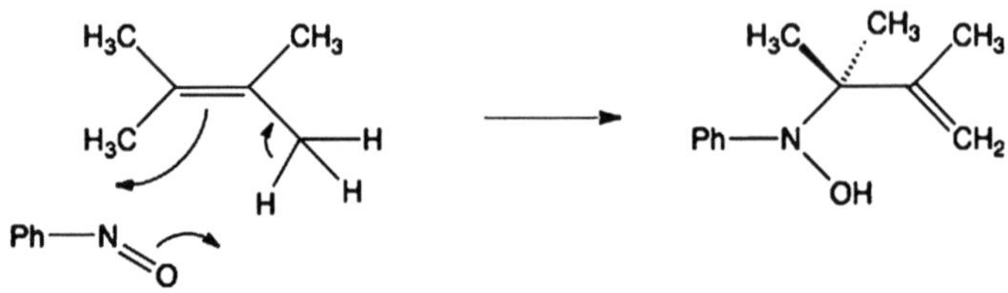

Scheme 5.2

Scheme 5.3

The reaction mechanism of the adduct formation, proposed by Sullivan, can be explained as a pseudo Diels-Alder reaction of nitroso group with the alkene molecule (Scheme 5.2).

Oxidation of the unstable hydroxylamine product by unreacted nitrosobenzene or oxygen yields corresponding nitroxide radical (Scheme 5.3).

However, the uncertainty in clarifying the *in vivo* processes induced by the nitrosophenyl compounds appears from the possibility of formation of fatty acid radicals by quite another mechanism. Since such a preexisted fatty acid radical can just be trapped by the injected nitrosoaromate, the observed nitroxide radical – fatty acid adduct, may be result of this secondary trapping process. For instance, the 12,13-epoxylinoleic acid radical can be formed by iron ion mediated lipoxidase oxidation of linoleic acid, which than can be trapped by the nitrosobenzene derivative [8].

Synthetic studies have shown that nitrosobenzene can react with steroids giving the nitroxide radical. In reaction with cholesterol, the hydroxylamine derivative is converted to corresponding radical (Scheme 5.4) [9].

5.2 Reactions with Thiol Consisting Biomolecules

Biomolecules, which have cysteine residue are possible targets for nitroso compounds because the cysteine RSH group, or its conjugated base, RS^-, behave as strong nucleophiles. Consequently, cysteine residues in hemoglobin, myoglobin,

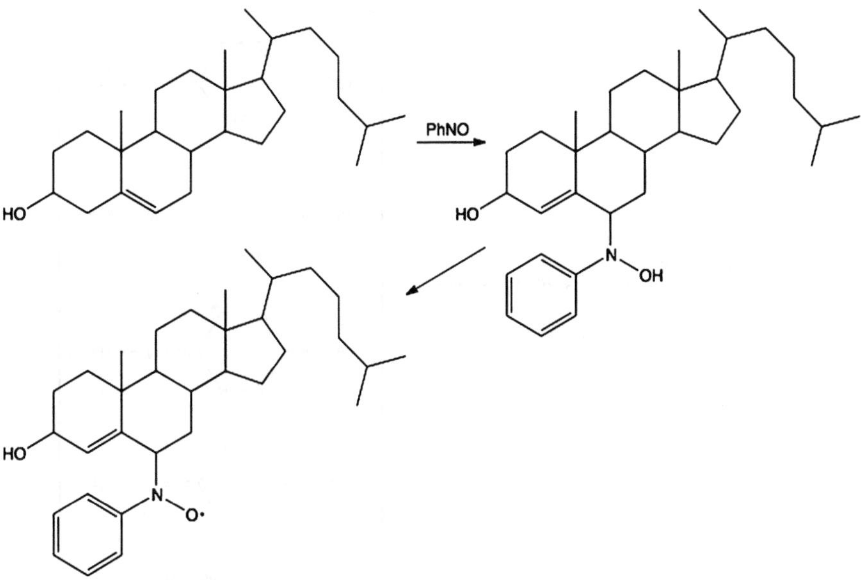

Scheme 5.4

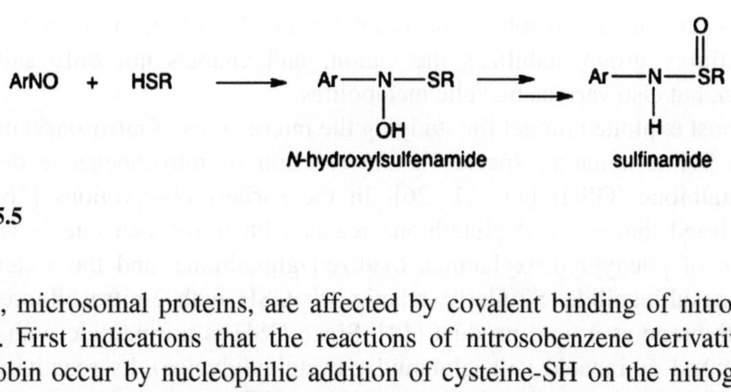

Scheme 5.5

albumin, microsomal proteins, are affected by covalent binding of nitrosoarenes [10–19]. First indications that the reactions of nitrosobenzene derivatives with hemoglobin occur by nucleophilic addition of cysteine-SH on the nitrogen atom of the NO group with the formation of the semimercaptal-like intermediate, have been reported by Eyer and Ascherl [20]. Nitrosoaryl compounds undergo complex reactions with thiols leading to the formation of *N*-hydroxysulfenamide [14–16, 21, 22] that is finally transformed to the corresponding sulfinamide (Scheme 5.5) [23].

The details of these rearrangements were studied by classical methods of physical organic chemistry and also by instrumental methods such as fast-atom bombardment mass spectrometry [24].

Obviously, the formation of *S*-conjugates with the cysteine residues of biomolecules occurs through the sulfenamide cation (Scheme 5.6) as an intermediate [25]. Its role in the mechanism has been investigated by using the

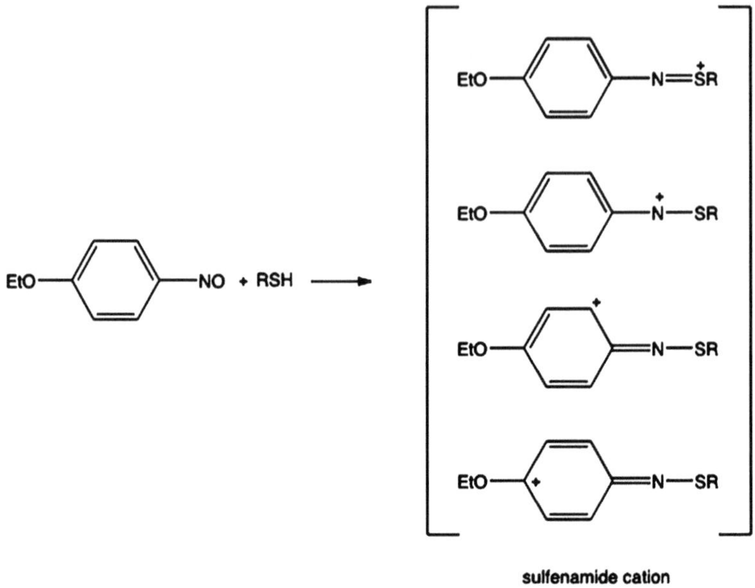

Scheme 5.6

p-ethoxy-substituted nitrosobenzene as a substrate. The electron-donating property of the ethoxy group stabilizes the cation, and enables not only sulfinamide formation, but also various bicyclic metabolites.

The most exploited model for studying the interactions of nitrosoaryl molecules with the SH terminating species is the reaction of nitrosobenzene derivatives with glutathione (GSH) [11, 21, 26]. In the earliest observations [26], it has been reported that reduced glutathione reacts with nitrosobenzene *in vitro* with formation of phenylhydroxylamine, oxidized glutathione, and the water-soluble glutathionesulfinanilide. Similarly, reaction of GSH with 2-nitrosofluorene gives 2-aminofluorene as a final product [14]. Since GSH is a tripeptide with cysteine as the central aminoacid (γ-L-glutamyl-L-cysteinylglycine), its reactivity toward nitrosoaryl derivatives is based on the addition of SH group to the nitroso double bond to give semimercaptal (Scheme 5.7).

The mechanism resembles the addition of a thiol anion to the carbonyl group. Analogous reactions are described in the Chap. 3 of this book. It has been reported that the semimercaptal intermediate can be transformed either to sulfinanilide or to *N*-phenylhydroxylamine (Scheme 5.8) [21].

It has been proposed that the rearrangement to the sulfinanilide occurs *via* formation of the nitrenium cation intermediate (Scheme 5.9) [26]. The indication about the cationic nature of the intermediate is supported by the positive Hammett ρ^{+} constant for the reaction of differently substituted nitrosobenzenes.

The ^{18}O-tracer experiment has shown that the oxygen atom in this mechanism is not derived from the nitrosobenzene reactant [26]. Conversion to hydroxylamine

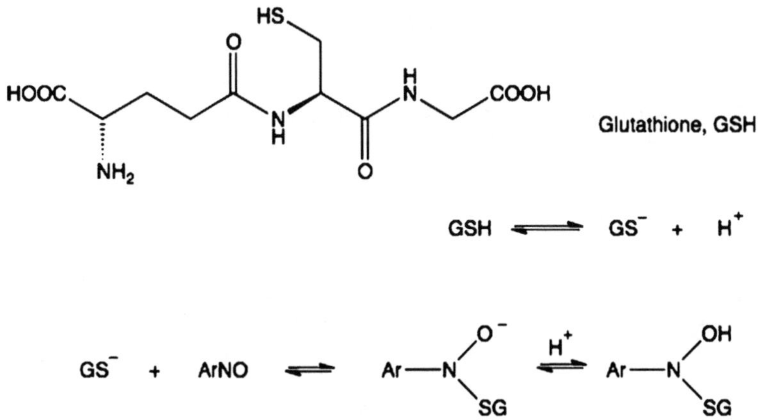

Scheme 5.7

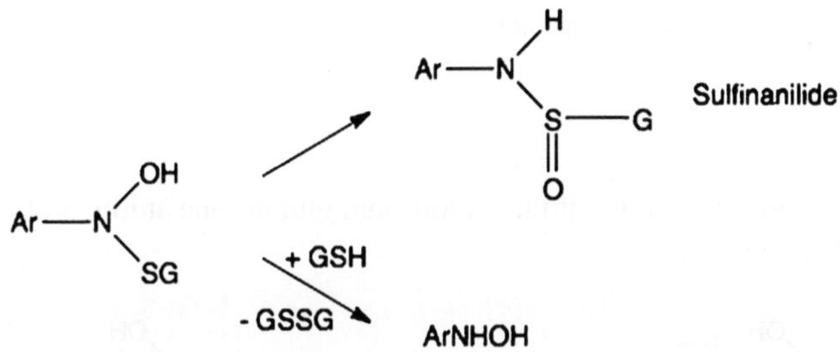

Scheme 5.8

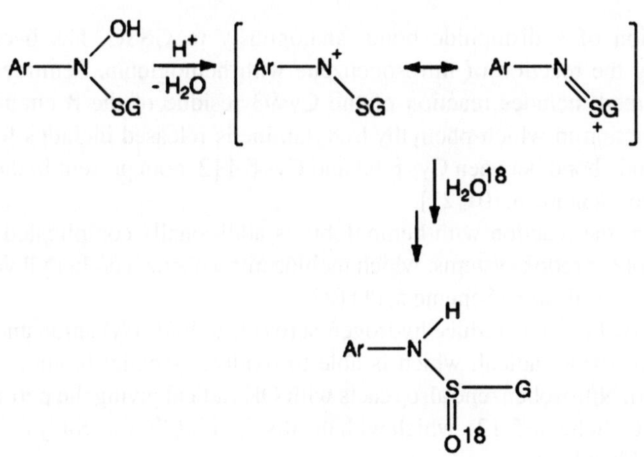

Scheme 5.9

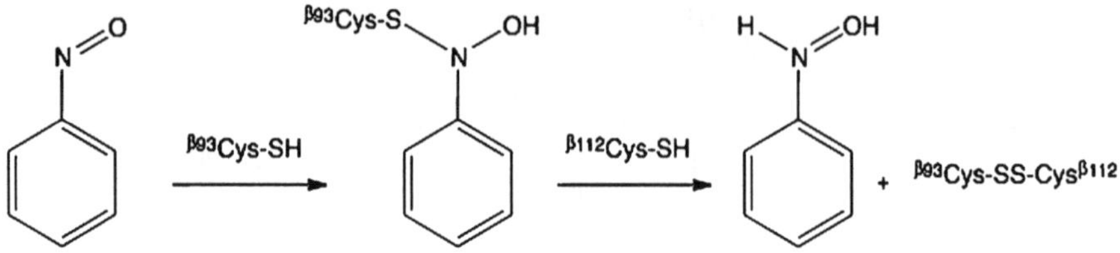

Scheme 5.10

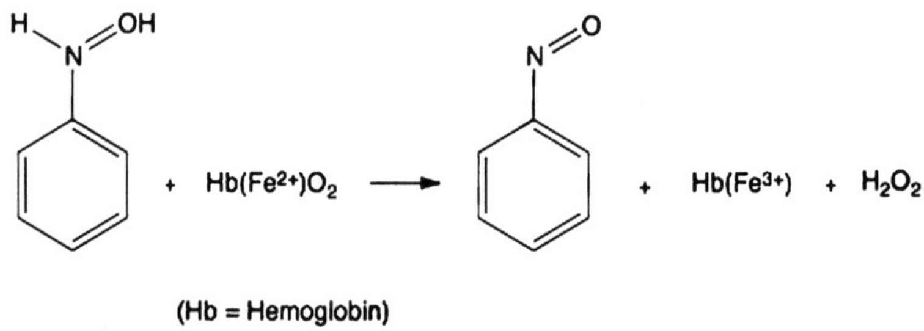

Scheme 5.11

derivative includes reaction with the additional glutathione anion, and formation of the disulfide by product:

Formation of a disulphide bond, analogously to GSSG, has been observed also during the reaction of nitrosobenzene with hemoglobin. Semimercaptal that is first formed includes reaction of the Cys93 residue of the β chain of globin. Further reaction in which phenylhydroxylamine is released includes formation of the disulphide bond between Cys β 93 and Cys β 112, both present in the β chain of hemoglobin (Scheme 5.10) [21].

However, the reaction with hemoglobin is additionally complicated because of formation of the redox systems, which include nitrosobenzene/phenylhydroxide and Fe^{2+}/Fe^{3+} equilibrium (Scheme 5.11) [27].

The cation Fe^{2+} can reduce hydrogen peroxide to hydroxyl anion and the highly reactive hydroxide radical, which is able to oxidize essential biomolecules inside the cell [28]. Nitrosobenzene also reacts with OH radical giving the peroxide radical intermediate (Scheme 5.12), which with the disulphide $^{\beta 93}$Cys-SS-Cys$^{\beta 112}$ yields the peroxide adduct [29].

Toxic nitrosoarene intermediates, which appear in living systems mostly on the pathway of the oxidation of arylamines also act as inhibitors of the enzymes included in this oxidation, the human arylamine *N*-acetyltransferases NAT1 and

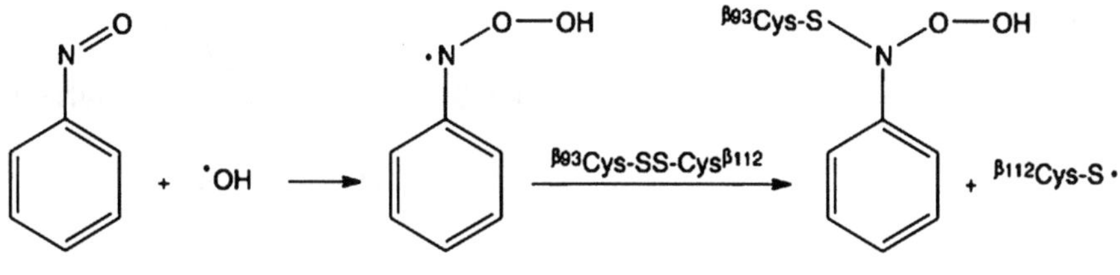

Scheme 5.12

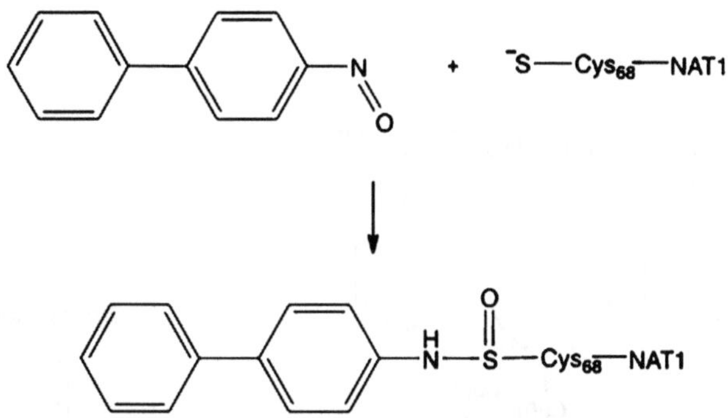

Scheme 5.13

NAT2. Hanna et al. [30] have reported the kinetic studies of the reactions of toxic 4-nitrosobiphenyl and 2-nitrosofluorene with NAT enzymes. They have found that, analogously to the previous examples, the nitroso molecule adds to the S-atom of the Cys68 residue of the NAT1 enzyme (Scheme 5.13). This example demonstrates that nitrosoaromatic molecule in the living systems could have double role, it can act as a toxic substance, but also contributes in detoxication.

5.3 Redox Reactions and Nitrosoaryl Compounds in Biological Systems

Aromatic nitroso compounds appear as intermediates in reductions of corresponding nitro compounds and oxidation of amines. Toxicity of nitro- and aminoarenes is attributed to their transformation to nitroso and hydroxylamine compounds in living systems [31, 32]. The reaction paths leading to the nitrosoaryl molecules in the first place include standard biological oxido-reductive equilibria such as NAD(P)/NAD(P)H systems, or ascorbic acid [33]. Besides the described reduction mechanisms by thiol-consisting biomolecules, nitrosobenzene can be reduced to phenylhydroxylamine radical by a one-electron reduction with NADPH *in vitro* [34]. By EPR spectroscopic investigation of the behavior of nitroso compounds in blood, Fujii et al. [35] have found that the corresponding hydronitroxide radicals

Scheme 5.14

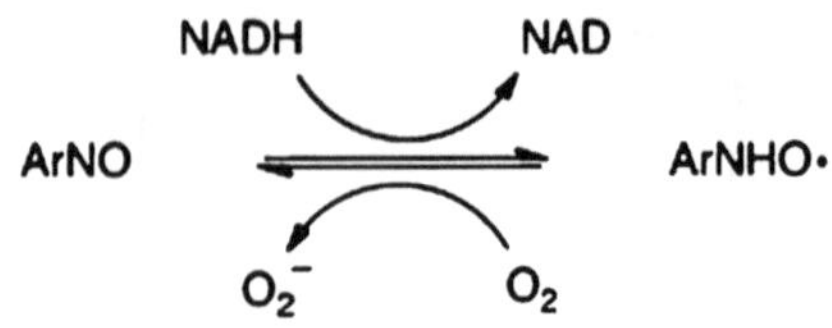

Scheme 5.15

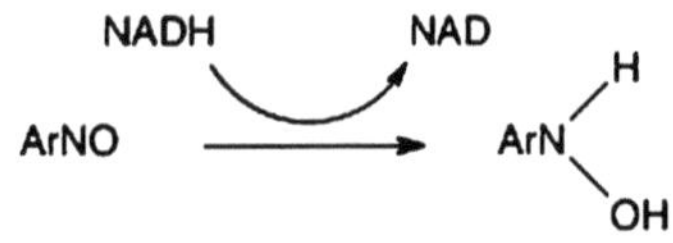

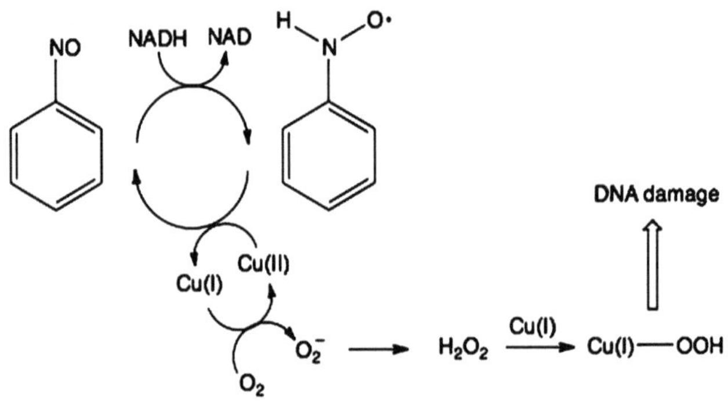

Scheme 5.16

are formed, followed by generation of superoxide anion. Under aerobic conditions, formation of both radicals are shown in the Scheme 5.14.

However, under anaerobic conditions, ArNO is directly reduced to hydroxylamine ArNHOH *via* two-electron reduction without the formation of hydronitroxide radical (Scheme 5.15).

Non-enzymatic reduction of nitrosobenzene by NADH seems also to be the basic step in the DNA damage caused by aryl nitroso compounds. Nitrosobenzene caused the formation of 8-oxo-7,8-dihydro-2′-deoxyguanosine in calf thymus DNA only if NADH and Cu(II) are present. As we have shown, nitrosobenzene is by NADH reduced to phenylhydronitroxide radical. Reoxidation to the starting nitroso compound can occur by electron from the Cu(II) → Cu(I) reduction process, which is additionally coupled with the formation of peroxide anion and hydrogen peroxide. Direct cause of the DNA damage is its reaction with the Cu(I) peroxide. All the reaction systems are shown in the Scheme 5.16 [36].

Copper peroxide generates hydroxyl radical that seems to be direct cause for the DNA single strand breaks. Let us remind that the formation of active OH radical is the process that can start with thiols as nitroso oxidants in the first step. Detailed study of the DNA cleavage induced by the thiol nitrosoaromatic compounds has been reported for the differently substituted nitrosobenzenes [37].

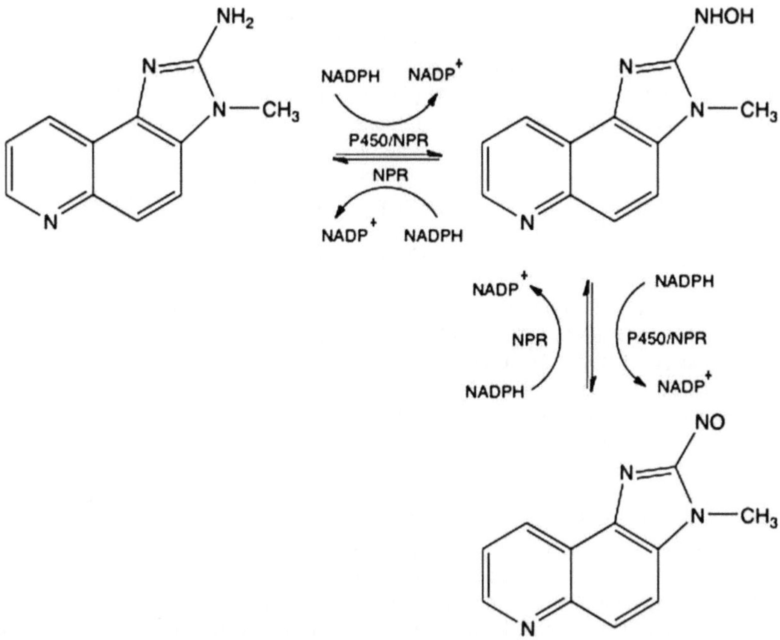

Scheme 5.17

Heteroaromatic nitroso compounds, which could induce mutagenic and cancerogenic effects appear in biological systems as products of partial oxidation of heterocyclic aromatic amines (HAA). Oxidation is catalysed by cytochrome P450 1A2 in combination with HADPH-P450 reductase (NPR). From the detailed kinetic studies of oxidation of the quinoline derivative, 2-amino-3-methylimidazo[4,5-*f*]quinoline, it follows that the reaction proceeds autocatalytically. The proposed autocatalytic system includes 16 reaction steps.

From the conceptual reaction scheme (represented below) is clear that the mutagenic effect of HAA can be diminished by action of NPR (Scheme 5.17) [38].

References

1. Miller JA (1970) Cancer Res 30:559–576
2. Saito K, Yamazoe Y, Kamataki T, Kato R (1983) Carcinogenesis 4:1547–1550
3. Mulder GJ, Kadlubar FF, Mays JB, Hinson JA (1984) Mol Pharmacol 26:342–347
4. Goodall CM, Moore CM, Stephens OB (1986) Xenobiotica 16:587–593
5. Eyer P, Kampfmeyer H, Maister H, Rosch-Oehme E (1980) Xenobiotica 10:499–516
6. Eyer P, Kampfmeyer H (1982) Chem Biol Interact 42:209–223
7. Fujii H, Zhao B, Koscielniak J, Berliner LJ (1994) Magn Res Med 31:77–80
8. Iwahashi H, Parker CE, Mason RP, Tomer KB (1991) Biochem J 276:447–453
9. Sridhar R, Floyd RA (1982) Can J Chem 60:1574–1576

10. Ascherl M, Eyer P, Kampffmeyer H (1985) Biochem Pharmacol 34:3755–3763
11. Eyer P, Lierheimer E (1980) Xenobiotica 10:517–526
12. Diepold P, Eyer P, Kampffmeyer H, Reinhardt K (1982) In: Snyder R, Parke VD, Kocsis JJ, Jollow DJ, Gibson CG, Witmer CM (eds) Biological reactive intermediates-II. Chemical mechanism and biological effects. Plenum Press, New York, pp 1173–1181
13. Eyer P, Schneller M (1983) Biochem Pharmacol 32:1029–1036
14. Dölle B, Töpner W, Neumann HG (1980) Xenobiotica 10:527–536
15. Mulder GJ, Unruh LE, Evans FE, Ketterer B, Kadlubar F (1982) Chem-Biol Interact 39:111–127
16. Klehr H, Eyer P, Schäfer W (1985) Biol Chem Hoppe-Seyler 366:755–760
17. Neumann HC (1984) Arch Toxicol 57:1–5
18. Lewalter J, Korallus U (1985) Int Arch Occup Environ Health 56:179–196
19. Hinson J, Mays JB (1986) J Pharmacol Exp Ther 238:106–112
20. Eyer P, Ascherl M (1987) Biol Chem Hoppe-Seyler 368:285–294
21. Eyer P (1979) Chem-Biol Interact 24:227
22. Saito K, Kato R (1984) Biochem Biophys Res Commun 124:1
23. Zuman P, Shah B (1994) Chem Rev 94:1621–1641
24. Montanari S, O'Carroll F, Paradisi C, Scorrano G (1995) Rapid Commun Mass Spectrom 9:1081–1082
25. Galleman D, Eyer P (1994) Environ Health Perspect 102:137–142
26. Kazanis S, McClelland RA (1992) J Am Chem Soc 114:3052–3059
27. Hirota K, Itano HA, Vedvick TS (1978) Biochem J 15:693–697
28. Halliwel B, Gutteridge JM (1990) Methods Enzymol 186:1–85
29. Di Girolamo F, Campanella L, Samperi R, Bachi A (2009) Ecotoxicol Environ Safety 72:1601–1608
30. Liu L, Wagner CR, Hanna PE (2008) Chem Res Toxicol 21:2005–2016
31. Khan MF, Wu X, Ansari GAS (2000) J Toxicol Environ Health A 60:263–273
32. Wulferink M, González J, Goebel C, Gleichmann E (2001) Chem Res Toxicol 14:389–397
33. Barkhan P, Howard AN (1958) Biochem J 70:163–168
34. Takahashi N, Fischer V, Schreiber J, Mason RP (1988) Free Radic Res Commun 4:351–358
35. Fujii H, Koscielniak J, Kakinuma K, Berliner LJ (1994) Free Radic Res 21:235–243
36. Ohkuma Y, Kawanishi S (1999) Biochem Biophys Res Commun 257:555–560
37. Hiramoto K, Ojima N, Kikugawa K (1997) Free Radic Res 27:409–418
38. Kim D, Kadlubar FF, Teitel CH, Guengerich FP (2004) Chem Res Toxicol 17:529–536

Index

H. Vančik, *Aromatic C-nitroso Compounds*, DOI 10.1007/978-94-007-6337-1,

MIX
Papier aus verantwortungsvollen Quellen
Paper from responsible sources
FSC® C105338

If you have any concerns about our products,
you can contact us on
ProductSafety@springernature.com

In case Publisher is established outside the EU,
the EU authorized representative is:
Springer Nature Customer Service Center GmbH
Europaplatz 3, 69115 Heidelberg, Germany

Printed by Libri Plureos GmbH
in Hamburg, Germany